Start Your Engines

Famous Firsts in the History of NASCAR

Jay W. Pennell

Foreword by Jeff Gluck

SPORTS
PUBLISHING

Sports Publishing books may be purchased in bulk at special discounts for sales promotion, corporate gifts, fund-raising, or educational purposes. Special editions can also be created to specifications. For details, contact the Special Sales Department, Sports Publishing, 307 West 36th Street, 11th Floor, New York, NY 10018 or sportspubbooks@skyhorsepublishing.com.

Sports Publishing® is a registered trademark of Skyhorse Publishing, Inc.®, a Delaware corporation.

Visit our website at www.sportspubbooks.com.

10 9 8 7 6 5 4 3 2 1

Library of Congress Cataloging-in-Publication Data is available on file.

Cover design by Tom Lau
Cover photo credit: AP Images

ISBN: 978-1-68358-470-4
Ebook ISBN: 978-1-61321-844-0

Printed in the United States of America

For Grandmom and Aunt Joyce.
I will forever be your racing buddy.

For kiddo. Thank you for putting up
with this crazy lifestyle and for always
being so understanding.

TABLE OF CONTENTS

Foreword

IMAGINE WALKING into your local Ben & Jerry's for a tasty cone of Cherry Garcia and being served by a NASCAR beat writer.

Few would ever guess the guy behind the counter scooping ice cream had been furiously pounding out a NASCAR news story or column just seconds before a customer walked in. But for several years, that was Jay Pennell's reality.

The author of this book is now a respected NASCAR writer, but nothing about that journey was easy. His path was completely self-made, forged by a relentless determination to chase a dream with a tireless work ethic that had him publishing stories in between passing out little spoonfuls of ice cream samples.

It's entirely appropriate that Jay is the one to write a book about NASCAR firsts, because he accomplished his own history in the sport. After all, Jay was the first "citizen journalist" to really make it in the NASCAR media.

Prior to Jay, the traditional path to becoming a NASCAR writer went something like this: Attend a journalism school, land a newspaper job, and work your way up the ladder. That was really the only way to the top.

And it's not like Jay didn't want to go that route. He worked for Street & Smith's on its *SportsBusiness Resource Guide & Fact Book*, hoping a foot in the door would land him a gig at the company's *NASCAR Scene* magazine. When I was on staff there, we'd get weekly race recaps e-mailed from Jay——unsolicited. But he never got the chance to really show what he could do.

So he struck out on his own. First, with a website called HardcoreRaceFans.com, Jay spent nearly three years building a presence from the ground up. He took advantage of NASCAR's "Citizen Journalist Media Corps," a program which invited bloggers and other writers into the media center to work alongside the regulars who covered races every week thanks to healthy travel budgets.

From the start, Jay's passion and knowledge for the sport stuck out. I remember when the beat writers would talk amongst themselves and say: "Hey, have you noticed that Jay kid? He asks really good questions."

And he did. The way he conducted himself earned respect from both media and drivers, and he eventually moved to sites called AllLeftTurns.com and Frontstretch.com, where he worked for two years.

Now you might be saying to yourself: I'm not familiar with those sites. Did they pay the bills?

No, they didn't——hence the *citizen* journalist and not *professional* journalist. Jay may have been just as professional as other writers, but he wasn't paid like one. That's where Ben & Jerry's came in. While most NASCAR writers would write their stories and call it a day, Jay maintained a heavy workflow while still working at his day job.

The stress of that juggling act must have been intense. I saw firsthand how difficult it could be when I asked Jay to join me as a writer for SBNation.com. Jay was crucial in helping get that site's NASCAR coverage on track (lame pun intended), but he had to do so while figuring out how to deal with more on his plate than any other writer.

But he was successful all the while, his role becoming increasingly prominent in the media——eventually rising all the way to FOXSports.com.

Like other pioneers, though, not everyone was always accepting of Jay. There were some media members who looked down on citizen journalists in general and scoffed at their presence.

What are they doing in here? They're just fans.

It would be easy to feel discouraged about attitudes like that. But through it all, I never saw Jay treat anyone poorly; he just put his head down and went to work.

Jay's journey to being the first citizen journalist to land a full-time gig in NASCAR came with probably more setbacks and false starts than anyone will ever know. This is a guy who would sleep in his car at the racetrack——as a habit, not just once or twice——in order to cover the sport he loved. The story might have a happy ending, but there was nothing simple about how he got there.

That's how trends get started and records get set, though——a theme I'm sure we'll see as we accompany Jay through this journey of NASCAR firsts.

Today, there's nothing unusual about sitting alongside a citizen journalist in a NASCAR media center. In fact, there are probably more of them on an average weekend than there are full-time reporters.

In some ways, though, they're all following Jay. He showed that it's possible to make it on pure grit and talent alone, whether anyone gives someone a chance or not.

How did he accomplish this? By making huge sacrifices and doing whatever it took to scratch and claw his way into the position he's reached today.

It's inspiring to look at Jay and remind ourselves that anything really is possible—and not just in terms of being in the NASCAR media. If you want something badly enough, like Jay did, you can make it happen.

Jeff Gluck, *The Athletic*

Introduction

THERE HAVE been few things in my life that have been more of a constant than NASCAR racing. From my earliest memories, to family vacations, to the career I now enjoy, NASCAR stock car racing has been a prominent factor since I was about three years old.

Growing up in New Jersey, I always felt I was the odd man out for enjoying NASCAR more so than the Philadelphia Eagles, Phillies, and Flyers combined. The fast-paced action, door-to-door racing and dramatic crashes were enough to capture my attention and passion from the earliest days.

My mother's side of the family raced throughout Alabama and Florida starting in the 1940s, primarily at Mobile International Speedway. My mother's Uncle Coolie Barnett, his son, Charlie, and their fast blue No. 62 were cemented into my head as lore and tall tales, yet they were true—for the most part.

Family reunions and other vacations often took our "Yankee" family down to see the "Rebels" in Mobile. After they made fun of our accents, told us how great the South was, and stuffed us full of food, we often headed to the racetrack to watch the Barnett Racing team beat up on the competition.

The first race I ever attended was a local short track event at one of the country's most historic tracks, Five Flags Speedway in Pensacola, Florida. Sitting with my family, I remember to this day watching two cars tangle on the backstretch and fly over the banking of the third corner. Five Flags Speedway did not have a retaining wall around the majority of the track, and when cars wrecked in the corners they often slid up and over the banking. To my young eyes, these cars flew off into an abyss and might never return. When both drivers emerged uninjured, I was immediately hooked.

After that first race, life would never be the same. As a child, I would religiously follow NASCAR and learn all I could about the sport, the drivers, the tracks, the history, anything I could wrap my head around.

On September 15, 1991, I would experience my first ever NASCAR race in person. Waking up in the predawn hours, the family made the nearly two-hour drive from Delanco, New Jersey, to Dover Downs, the Monster Mile. Sitting in the front-stretch grandstands that day, we watched Harry Gant stretch his fuel to win the race, his third of four consecutive wins that month, earning him the title, "Mr. September."

I would continue to attend NASCAR races nearly every year afterward, with our family often taking vacations simply to camp at a race. Our travels took us to Bristol, Tennessee; Charlotte, North Carolina; Watkins Glen, New York; and eventually to Daytona Beach, Florida.

My parents, Cindy and Jay, as well as my Aunt Joyce, were instrumental in fueling my passion for the sport. Whether it was the numerous trips to the racetrack, or the copious amount of money spent on merchandise, books, posters, and more, my family did all they could to foster my love of NASCAR.

As I grew older, the life of a teenager would often inter-rupt my passion for NASCAR. However, every weekend I kept up-to-date with the latest news and rumors from the track. I fervently recorded every race on VHS tape, and still have most of that collection.

During my senior year of high school, my Aunt Joyce surprised me with tickets to the 2002 Daytona 500. After years of going to races, I finally had the chance to see the "Great American Race" in person. This was a big deal. There was only one snag. My high school wrestling playoffs were scheduled for the same weekend.

As it worked out, the overall outcome of our team's advancement in the playoffs came down to my match. If I won, we moved on to the next round of the playoffs. If I lost, I was able to go to the Daytona 500 with no strings attached.

I love racing, but I would never throw away something I had worked so hard on, not even for the Daytona 500. Still, the thoughts of my dilemma were running through my mind heading into the match. I gave it my all, left it all on the mat, but lost. Struggling with the disappointment, I was able to hold my head high knowing my teammates saw I did my best to win the match. My wrestling career may have been over, but I was taking another step in my racing career.

When it came time to look at colleges, I only did so in North Carolina. While I was not going to school specifically for racing, I knew if I would ultimately want to get involved in NASCAR in some way, I would have to be in North Carolina, more specifically Charlotte.

So, that's what I did. I went to Queens University of Charlotte and was a double major in history and American Studies, receiving bachelor's degrees for both. While my passion for NASCAR waned a bit during my college years, it was never gone. Each Sunday you could find me watching the race intently, and my friends all suddenly knew more about NASCAR than they ever had—whether they like it or not.

Once out of college, I was stuck doing what many college graduates do—work in the foodservice industry and try to pay the bills. Working multiple jobs, I became a paid intern with Street & Smith's *SportsBusiness Journal* working on the annual *Resource Guide and Fact Book*. Certainly a step up from scooping ice cream and bussing tables, the opportunity to work for SBJ did more to further my career than I gave credit at the time.

Not only was I working in a functioning newsroom, I was also on the same floor, sharing office space with *NASCAR Scene* and *NASCAR Illustrated*, two publications I grew up reading and one day hoped to write for. One of the writers I admired most, Steve Waid, worked down the hall and was always welcoming when I sought advice and had questions.

At this time, social media was first starting to emerge. During my college years, I had developed a personal MySpace page, but now decided to make a racing-specific page. On there, I would write race recaps and anything that would come to mind. I had honed my writing skills in college covering topics such as German history and the historical relevance of American sub-cultures, but writing about racing was much different and took practice.

A friend from college saw one of my blog entries on MySpace about a rain-shortened race won by Jeff Gordon at Pocono Raceway. He worked for the *Laurinburg Exchange* out of Laurinburg, North Carolina, and asked if he could use the race recap in the paper, which of course I agreed to. On June 13, 2007, my byline first appeared in an actual news outlet as my article titled, "Mother Nature Lends Father-to-be a Helping Hand," appeared next to two stories from the *Associated Press*.

To help get my name out there and distribute this new passion for writing about NASCAR, I took full advantage of the fact I had Street & Smith's company e-mail list. Once I published a blog on the MySpace page, I would email the link to everyone on the *NASCAR Scene* and *NASCAR Illustrated* staff.

Eventually, I would seek advice from those other than Waid. Yet, this was still the height of NASCAR's media days and someone with no real print journalism experience just wasn't going to cut it. If I wanted to be part of the NASCAR media world, I would have to find a different way in.

That door opened on a blistering cold day in January 2007 at the groundbreaking of the NASCAR Hall of Fame in Charlotte, North Carolina. Sitting in the makeshift tent as some of NASCAR's most iconic participants broke ground, I took notes for a blog post about the historic day. After the pomp and circumstance, I was met by a gentleman by the name of Ray Everett. He was in the process of creating a new NASCAR news website and was looking for writers.

After a few meetings, I went to work helping Everett create and run HardcoreRaceFans.com. This was my first venture in online content management, writing about NASCAR and so much more. The entire process from start to finish was a learning experience—both good and bad. Yet this is where I honed my skills and learned all I could about how to better tell the story of NASCAR to the readers.

Once I parted ways with Everett after nearly three years, I went to work for a myriad of websites including Frontstretch, All Left Turns, SB Nation, and Athlon Sports. Nothing was easy, and few things were given to me. Each step I took along the path—and continue to take along that path—has made me a stronger person, better writer, and more effective NASCAR journalist.

Currently, I serve as a digital content writer and editor at FOX Sports. Working with award-winning journalist Tom Jensen, I continue to learn every day, not only about the sport but also about the business behind it, the work it takes behind the scenes, and many other things. I am now a colleague of some of the competitors I watched growing up, people like Darrell Waltrip, Larry McReynolds, Jeff Hammond, and starting in 2016, Jeff Gordon. My career path has led to my work being published in *NASCAR Illustrated*, *2012 Athlon Sports NASCAR Racing Preview*, and even Playboy.com.

My journey to writing this book is long, winding, and full of stories, but that could take up another book entirely. This is a book about firsts, and it is fitting that this is my first book. This process has been an eye-opening experience and one I will never forget.

I certainly would not have this opportunity if it were not for Ben White. Ben was one of those writers I thoroughly enjoyed reading as a kid. When I fought my parents about reading, we would often compromise on NASCAR-related topics. Ben was often the man behind those words. We have grown to know one another over the years, and he was instrumental in helping put this together. I will forever be thankful and honored both for his confidence and support.

My wife, Courtney, and daughter, Abigail, have been my biggest champions through the process of writing this book. They were there when things became stressful, they gave me the space I needed to focus solely on this project, and they were my motivation for completing the task.

I also have to thank my sister, Nicole, and brother, Travis, neither of whom are race fans. They were often dragged along to the racetrack on family vacations and probably know more about the sport than they ever wanted to. Nicole has always helped me become a better writer, and she played a crucial role in steering my career toward journalism, whether she knows it or not.

I could not do this without the constant support of my family. They have put up with my love for racing my entire life and have gone above and beyond to accommodate that passion. I am eternally grateful.

This book is not a complete list of NASCAR firsts, but instead highlights some of the most important moments and events in the sport's storied history. The final product is one I put a lot of blood, sweat, and tears into, and I hope you enjoy reading it.

Jay Walter Pennell
June 18, 2015

Introduction to the Paperback Edition

As I sit here and complete the final edits on the paperback edition of *Start Your Engines: Famous Firsts in the History of NASCAR*, I do so on a truly historic weekend for the sport covered in this book.

NASCAR is in the middle of its 75th anniversary season, the sport's 75 greatest drivers were recently celebrated at Darlington Raceway on Throwback weekend, and the annual NASCAR All-Star Race was just held at North Wilkesboro Speedway.

This marked the first time NASCAR returned to the once abandoned and deteriorating speedway since 1996, something few in the racing world thought they would ever see happen. Sitting outside the track in a makeshift press box—which was really a few tables and chairs on a riser under a tent—and looking around the historic facility that traded weeds and crumbling infrastructure for tens of thousands of avid race fans and a completely renewed sense of life, it was amazing to see the sport thriving but recognizing its amazing history and past.

Since the first edition of this book was published in 2015, there have been a lot of changes and historic firsts for NASCAR. From a new ownership model, to a brand-new race car, to a season without fans, to the first-ever NASCAR street race, the sport founded in 1948 is still alive and well, thriving, pushing to the future, but also honoring its heritage.

Things have also greatly changed in my career in motorsports and the journey from where I was when the book was first sent to publication in 2015 to this second edition in 2023 has been long, winding, memorable, and something I will never forget.

Shortly after publication, my career took a turn as I left FOX Sports and the writing world behind for a while. I joined Richard Childress Racing as Manager, Content and Communications for four seasons. Working for one of the most iconic teams in motorsports was a dream come true—even if I was a Rusty Wallace fan growing up. In short, I handled public relations for drivers such as Daniel Hemric, Paul Menard and Brandon Jones, while working with some of NASCAR's best behind the wheel and in the garage. I helped plan RCR's 50th anniversary content in 2019 and tried to bring as much of the organization's history to life for the fans.

On February 18, 2018, on the 20th anniversary of Dale Earnhardt's first Daytona 500 win and on the 17th anniversary of his tragic death on the final lap of that race, Austin Dillon and the No. 3 Richard Childress Racing team made a dramatic last-lap pass after contact with Aric Almirola to win the organization's third 'Great American Race.'

That day, I handled the RCR Instagram account and captured behind-the-scenes content from pit road, Victory Lane, and the celebration throughout the night. My role was to provide RCR fans the opportunity to experience the historic win virtually through social media and the camera on my phone.

Standing in Daytona Victory Lane with the No. 3 RCR Chevrolet still steaming and covered in confetti, surrounded by the entire team, Richard Childress standing proud of his grandson and his team, it was quite the moment to take in. All I could think of was the people who helped get me to this point. My parents, Aunt Joyce, Grandmom, Charlie Barnett.

The day was truly a dream come true for a kid that grew up absorbing all he could about NASCAR, just hoping one day he could be involved in the industry somehow. Today, those dreams have become a reality and continue to evolve and change.

This career has taken me places I never imagined, allowed me to meet my heroes and experience things I grew up just hoping to be a part of somehow. From writing, to working PR, content creation, to video games and virtual races, my career path since the first publication of this book has had more twists and turns than a NASCAR road course, but that's for an entirely different book.

I am grateful to still be involved in the sport that I grew up loving. I appreciate all the friends that I have made in the industry and everyone that has helped me along the way and continue to do so. I am also appreciative of my friends outside the industry for always allowing me to talk their ears off about NASCAR and for always being there when needed—Sin Miedo for life.

None of this would have been possible without the understanding of my daughter Abigail. She has endured my lengthy trips away from home, the long working hours, and more conversations about NASCAR than she could have ever wanted. She is an incredible person and what I am most proud of in this life.

This is an incredible opportunity, and I am thankful to so many that have helped me get here, especially Ben White, Jeff Gluck, Julie Ganz and Stephan Zguta.

I appreciate everyone who reads this updated version of *Start Your Engines: Famous Firsts in the History of NASCAR*, and I hope you learn something about the sport that so many have loved for seventy-five years. Here's to many more years of thrilling racing, close finishes, and famous firsts.

Jay Walter Pennell
May 22, 2023

The First Organized Meeting of NASCAR

PRIOR TO the formation of NASCAR, stock car racing in the United States was a hodgepodge of organizations and tracks, with no central sanctioning body and few consistent rules. Born out of the days of outrunning government revenuers and hauling moonshine, stock car racing gained popularity, particularly throughout the Southeast, in the 1930s and 1940s.

From Georgia, to Florida, to the Carolinas, stock cars could often be found racing in cow pastures, small clay and dirt tracks, and along the beaches of Daytona.

The American Automobile Association (AAA) was the dominant racing organization before NASCAR was created, hosting primarily open-wheel Indianapolis-style races, with the occasional stock car race featured as well. The AAA had abandoned stock car racing at the end of 1946, with their Contest Board declaring it a "fad . . . that is dying out."

By 1947, there were three prominent stock car racing organizations in the United States: the National Championship Stock Racing Association (NCSRA); the National Stock Car Racing Association (NSCRA or NSRA); and the National Championship Stock Car Circuit, owned by Bill France.

After a year of struggling to keep rules, drivers, and mechanics under control due to the assortment of sanctioning bodies, France decided to call together stock car racing's bigwigs at the end of 1947 to talk about the future of the sport.

France announced his invitation in racing publications such as *Speed Age*. His invitation garnered attention from racers up and down the East Coast. An estimated thirty-five drivers, team owners, promoters, and mechanics made the trip to the Streamline Hotel on Highway A1A in Daytona Beach, Florida, for a meeting of the minds that would forever change the course of racing history.

The group gathered in the Streamline Hotel's top floor Ebony Bar for the first time on December 14, 1947. Knowing the meetings would take quite some time and could become stressful, France brought in local girls to keep the gentleman entertained and at ease throughout the course of the three days.

France opened the meeting and set the tone by making it clear that stock car racing needed to be organized if it wanted to grow and flourish. "Nothing stands still in the world," France said. "Things get better or worse, bigger or smaller."

Always looking toward the future, France did his best to convince the group that the sport of stock car racing could see exponential growth if they could come together to create an organized sanctioning body with a legitimate championship point system.

"Stock car racing has got distinct possibilities for Sunday shows and we do not know how big it can be if it's handled properly," France told the group, proclaiming it could become as big as open-wheel racing. "I believe stock car racing can become a nationally recognized sport by having a national point standing. Stock car racing as we've been running it is not, in my opinion, the answer. . . . We must try to get track owners and promoters interested in building stock car racing up. We are all interested in one thing that is, improving present conditions. The answer lies in the group right here today."

The group spent three days debating and discussing the future of stock car racing, all while smoking, drinking, and enjoying the company of the ladies France brought in to provide entertainment.

As the men continued to go over the specifics of organizing the hodgepodge of racing associations into one cohesive group, issues such as the point system, racing divisions, and other details would be hashed out with the help of whiskey and cigars.

While France was able to gather the large group of stock car racing drivers, promoters, owners, and members of the media, not everyone was ready to buy into his idea fully.

"He had to be something special, 'cause I don't know of any racer that liked him, but we kept going back for more," legendary mechanic Smokey Yunick remembered. "France was a world-class bullshitter and had the balls of an elephant with regards to gambling with finances, and he'd work twenty hours a day, seven days a week if necessary."

After three days of meetings, the group had come together to determine the championship points system, three divisions of racing, and now was faced with the task of naming the newly formed racing organization.

"We had to come up with a name," France's partner Bill Tuthill said of the meeting. "Red Byron wanted National Stock Car Racing Association. Red Vogt suggested . . . National Association for Stock Car Auto Racing (NASCAR) . . . Then somebody pointed out that there was already an association in Georgia with the NSCRA name, so Ed Bruce moved to disregard the first name and incorporate, in Florida, under the name of National Association for Stock Car Auto Racing. Jack Peters seconded, and it became the official name of the new organization."

With an official name, point system, and racing divisions, the next step was to determine the leadership of the newly founded organization. Red Byron suggested France, the driving force behind the meeting, lead the charge as the president of NASCAR. Whether the group knew it or not when they parted ways from the three-day meeting at the Streamline Hotel, NASCAR would be run by the France family for the next seventy-five years.

While France would be tapped to run the new organization as president, E. G. "Cannonball" Baker would serve as the national commissioner, Eddie Bland as vice president, Bill Tuthill, secretary, and Marshall Teague, treasurer.

A prominent figure in the three-day meeting at the Streamline Hotel and one of the most savvy and successful mechanics in stock car racing's earliest days, Red Vogt felt abandoned by France when he was not included in the organization's decision-making group. In 1954, Vogt would be given the first lifetime NASCAR membership for his role in helping organize and name the sport during those three-day meetings.

Despite some hard feelings and hesitations over the outcome of the meeting, France remained vigilant in his quest to make NASCAR not only the premier stock car circuit, but the number one racing organization in the nation.

"The purpose of this association is to unite all of stock car racing under one set of rules; to set up a benevolent fund and national point standings system whereby only one stock car driver will be crowned national champion," said France. "Every track and every area has a 'national champion' of every type of racing. This has so confused the sportswriters that they give up in disgust after trying to give the public an accurate picture."

With NASCAR now on the scene, and France in charge, it would be clear to all which stock car racing circuit was number one in the US.

NASCAR's First Official Season

WALKING OUT of the Ebony Room of the Streamline Hotel on Highway A1A in Daytona Beach, Florida, after three days of meetings, Bill France had accomplished his goal of not only organizing a new national sanctioning body for stock car racing but also securing the majority of the control.

While the meeting ended with officials being nominated and the illusion of a democratic organization, NASCAR was truly owned by three stockholders. Louis Ossinsky, an old friend of France's and the lawyer hired to help incorporate NASCAR, owned 10 percent of the shares. Bill Tuthill, NASCAR's new secretary, owned 40 percent of the organization, leaving 50 percent of the shares to France, giving him total control of NASCAR.

"The next thing we know, NASCAR belongs to Bill France," said team owner and meeting attendee Raymond Parks.

France worked with Ossinsky to fill out the proper paperwork to have the newly formed NASCAR incorporated as an official organization. However, he would not wait for the paperwork to be filed to start NASCAR's march into the future.

Of the three divisions determined in the Streamline Hotel meetings, the Modified Division would take center stage in 1948. The modifieds had been the most popular form of stock car racing in the country at the time, dating back to pre–World War II days.

The Strictly Stock Division—the forerunner to what would become the NASCAR Cup Series—would not begin racing until the 1949 season. The idea of racing family sedans like those sold on showroom floors was something France hoped would catch on. While the postwar economy was starting to produce more family sedans, it would take some time for the public to accept the new cars beating and banging on the racetrack.

On January 4, 1948, a race was held at the one-and-a-quarter-mile Pompano Beach Speedway dirt track in Florida. While the race was not an official NASCAR event, it kicked off the 1948 season and featured a field of drivers committed to racing with France's newly formed sanctioning body. Buddy Shuman won the race.

The first officially sanctioned race for France's new NASCAR took place in his own backyard on February 15, 1948. A total of sixty-two cars entered the event that would take place on the Daytona Beach and Road Course, where France had promoted and raced for years.

Prior to the event, France and NASCAR National Commissioner E. G. "Cannonball" Baker honored drivers Fonty Flock and Ed Samples. Flock was the 1947 National Championship Stock Car Circuit champion, and Samples was the runner-up and 1946 champion. Before kicking off the new era of NASCAR, France made sure to recognize those that were successful in his previous series.

With 14,000 fans in attendance, the fifty-car field took the green flag on the road and beach course driving into history. Many drivers complained of the soft sand, and through attrition, mechanical failures and accidents, only twelve cars finished the race.

Red Byron won driving a Red Vogt–built and Raymond Parks–owned modified 1939 Ford. The World War II veteran cemented his place in history as the first NASCAR winner, but Byron was not done making history yet.

On February 21, 1948, the paperwork filed by France and Ossinsky became official, and NASCAR was incorporated.

Throughout that first season, NASCAR would host a total of fifty-two races, often running multiple races in multiple states on the same day. On May 23, 1948, NASCAR held simultaneous events in Macon, Georgia; Danville, Virginia; and Dover, New Jersey.

That first season was not without tragedy, either. As was often the case, NASCAR held two races on July 25, 1948, one in Columbus, Georgia and the other in Greensboro, North Carolina. After a strong and energetic start, NASCAR would be forced to deal with tragic situations at both tracks.

In Columbus, Red Byron was doing all he could to hold off a hard charge from Billy Carden and championship rival Fonty Flock in the closing laps. As he went to block a run in the corners, the right front tire exploded and sent Byron's Ford sliding up the small banking and through the chicken-wire fencing into the crowd.

One spectator, a forty-six-year-old man from Alabama, would have his lower leg amputated. A twenty-four-year-old woman suffered a broken pelvis. Three others suffered broken legs, while twelve people were treated for cuts, bruises, broken ribs, and other injuries.

Seven-year-old Roy Brannon was sitting on the hood of his parent's car when Byron slid into the crowd. One of the fence posts separating the crowds from the track snapped upon impact and struck the child in the head. His parents immediately took him to the hospital, where it was determined he had a skull fracture. The boy would succumb to his injuries.

A total of seventeen people were injured in the late-race incident, but the tragedy was not over yet. In the day's other race in Greensboro, Bill "Slick" Davis flipped his car and was thrown from the vehicle. Laying on the track, Davis was hit by his car when Tim Flock and three other drivers could not avoid his empty, wrecked car. Davis would die later that night of head injuries.

Shaken from the incident in Columbus, Byron was distraught but remained focused on beating rival Fonty Flock for the championship. However, the savvy stock car driver also had a dream of racing in the Indianapolis 500. Byron made an attempt to qualify for the 1948 running of the famed open-wheel AAA event at Indianapolis Motor Speedway, but when motor problems kept him off the track for multiple days, threatening his plan to run stock car races, Byron left Indy and returned to France and NASCAR to chase the championship.

Byron and Flock would exchange the points lead a total of five times over the course of the fifty-two races, with Byron winning eleven and Flock winning fifteen. When Byron won the season's final race, France determined he finished 32 points ahead of Flock and was named the 1948 champion.

After laying much of the groundwork for NASCAR, Byron, Vogt, and Parks were able to celebrate the honor of becoming the first championship team in NASCAR history.

The 1948 season was celebrated in Daytona Beach, Florida, in January 1949. Five thousand of the $64,000 NASCAR earned in ticket sales from the first season had been set aside and divided among the top twenty drivers in the season standings. NASCAR secretary Bill Tuthill signed each of the checks personally, with Byron taking home a total of $1,250. As part of his contract, Byron gave $834 of his earnings to car owner Raymond Parks.

Major changes would come to NASCAR between the 1948 and 1949 seasons, most notably the shift from the Modified Division to the Strictly Stock Division. While most look to 1949 as the first season of NASCAR's premier series, it officially kicked off with the fifty-two race modified schedule, bookended with wins by its first champion Red Byron.

The wheels were rolling, both literally and figuratively for NASCAR, and the sports world would never be the same.

The Strictly Stock
Division Is Born

AFTER A successful first season that saw the brand new NASCAR modified series host fifty-two events and feature a tight championship battle, Bill France was pleased with his creation, but knew more had to be done to set NASCAR apart.

While modifieds had been the most popular form of stock car racing for some time, France believed NASCAR needed something else. And so did his rivals.

After the historic three-day meeting at the Streamline Hotel in Daytona Beach, Florida, in December 1947, the majority of stock car racers agreed to follow France's lead and tow the NASCAR line. However, a number of other racing organizations still existed and threatened France's plan for supremacy.

Just an upstart organization with one official season under its belt, NASCAR and Bill France were now tasked with sanctioning nearly 400 races in 1949, with new facilities taking his version of stock car racing to new parts of the country. Through this expansion, France's NASCAR clashed with the other racing series, primarily O. Bruton Smith's National Stock Car Racing Association (NSCRA).

Stock car racing was not a fad, as the AAA had once called it, and people were flocking to races to experience the noise, speed, and danger in person. Some of France's drivers lobbied to have NASCAR change its policy on purse money, doing away with a guaranteed amount and instead paying 40 percent of the money made on ticket sales. When France refused to give in on the guaranteed purse money, a number of drivers defected for other series, primarily Smith's NSCRA.

To penalize drivers who defected to rival racing organizations, France and Bill Tuthill announced in 1949 that any NASCAR driver that raced in a competing series would lose all points accumulated to that point. Some tested the new policy and raced in a NSCRA event, and thirteen drivers lost all their NASCAR points.

Tensions were growing and France kept a close eye on Smith and his NSCRA. Smith was looking to take control of stock car racing, and France was doing all he could to maintain and ensure dominance. Both men were looking to the future.

Postwar production was now churning out more new automobiles than ever before, and the booming economy allowed everyday Americans to own a car. With that in mind, France believed a new strictly stock division would resonate with fans because the cars racing out on the track were the same ones being sold on the showroom floors. The idea was to keep the rules strict against any modifications, making the cars the same as those fans could buy the following day.

The idea of strictly stock racing was not an original idea. The AAA first sanctioned a strictly stock race in 1926, with the headline, "Will the Car Like You Drive Win Over the Kind Driven by Your Friend?" The AAA again sanctioned strictly stock races at the Daytona Beach and Road course in 1936, with France running the promotion. He even raced in a handful of strictly stock events during his time as a driver.

While the modified division helped to get NASCAR off running, France decided to experiment with his idea of a strictly stock division. NASCAR held its first Strictly Stock Late Model race on January 23, 1949, at the two-mile Broward Speedway in Florida, with Lloyd Christopher taking the checkered flag in the 10-mile event. The Strictly Stock race was not the main event of the day, though. That honor went to the 100-mile modified division race.

Wanting to give it a second try, NASCAR sanctioned another ten-mile Strictly Stock race on February 27, again at the Broward Speedway. The field soared around the two-mile speedway, with Benny Georgeson taking the checkered flag.

France and NASCAR were experimenting with Strictly Stock races, but they were both short events, neither the main attraction. In early 1949, France's rival, Smith's NSCRA, announced it would host a Strictly Stock race as its main event.

France knew he had to counter and protect NASCAR's dominance over the NSCRA. He did so in May, by announcing plans for NASCAR to host a $5,000, 200-lap Strictly Stock race in Charlotte, North Carolina on June 19, 1949.

France welcomed any driver that met the strict stock rules to race in the 150-mile event at the three-quarter mile Charlotte Speedway dirt track. To show his displeasure for their defections to the NSCRA, France did not allow Marshall Teague, Buddy Shuman, Speedy Thompson, or Ed Samples to compete in the NASCAR event. As the race neared, each driver returned to France looking for an opportunity to compete in the first-of-its-kind event. Hearings were held for each individual, and they were reinstated after paying fines ranging from $50 to $150.

After multiple days of practice and qualifying, race day finally rolled around and over 22,000 fans headed to the dirt track to check out NASCAR's newest form of racing. The field was full of some of the day's biggest stars: Red Byron, Curtis Turner, the Flock brothers, and female driver Sara Christian. The new

Strictly Stock rules also opened the field up to nine different manufacturers: Buick, Cadillac, Chrysler, Ford, Hudson, Kaiser, Lincoln, Mercury, and Oldsmobile.

Brothers Bob and Tim Flock led the 33-car field to the green flag to kick off a new era in motorsports history. Bob Flock led the opening five laps of the race from the pole but was forced out of the race early after an engine failure. He would substitute behind the wheel for Christian later in the event. With Bob Flock out of the running, Bill Blair's 1949 Lincoln took the lead and remained out front for 145 laps.

The race featured a spectacular crash on Lap 107 when Lee Petty flipped his Buick in the third turn. Petty, who would go on to become the patriarch of one of racing's biggest dynasties, was making his first NASCAR start and had, earlier in the week, driven the borrowed car down from Level Cross, North Carolina.

In front of the field, Blair set a blistering pace, lapping the competition. With roughly 50 laps to go, the radiator gave out on Blair's Lincoln, and he was forced to park the car. Blair's misfortune gave way to Glenn Dunaway, a driver who did not have a ride the morning of the race but was put in a 1947 Ford by Hubert Westmoreland.

Dunaway would lead the final 47 laps and take the checkered flag over Jim Roper to win NASCAR's first official Strictly Stock race. All Dunaway needed to do to earn the $2,000 purse and cement his name in history was pass NASCAR's teardown process by Technical Inspector Major Al Crisler, as questions had been raised about how smooth the car performed on the rough dirt track.

It was determined the Westmoreland-owned Ford had modified bootlegging-style springs in the rear of the car and was in violation of the Strictly Stock guidelines. NASCAR disqualified Dunaway's car, giving the win to Roper, who was three laps down. Fonty Flock was second, Red Byron third, Sam Rice fourth, and Tim Flock was fifth.

Westmoreland protested, filed a lawsuit, but ultimately NASCAR and Bill France would prevail.

A total of eight races were held during the 1949 Strictly Stock season, with Red Byron and Bob Flock each winning two events. The Strictly Stock division raced at tracks up and down the East Coast, as far south as the traditional Daytona Beach and Road Course and as far north as the Hamburg Speedway in New York.

After flipping his borrowed car in the Charlotte race, Petty would earn his first NASCAR victory on October 2, 1949, at Heidelberg Speedway in Pittsburgh, Pennsylvania—the first of many for the Petty name.

In the end, Byron's two wins, four top-fives and four top-tens were enough to give him an 117.5-point lead over Petty and the first Strictly Stock championship.

It would be the final season Byron would win NASCAR races and the last time he would compete on a full-time basis. During the 1950 season, Byron would run multiple non-NASCAR sanctioned events and was twice stripped of his points by Bill France. His time in France's NASCAR was coming to a close. Byron was a hard-nosed driver that helped pave the way in NASCAR's earliest days, earning championship titles in both 1948 and 1949.

Those eight races in 1949 laid the groundwork for something that would develop and evolve into one of the world's most successful sports and a multi-billion-dollar industry, spanning three-quarters of a century.

A Southern Tradition Begins: 500 Miles at Darlington

AFTER A trip to the 1948 Indianapolis 500, South Carolina peanut farmer Harold Brasington had a dream. He imagined a 500-mile race in the South that could one day rival the open-wheel spectacle that had so much history. The only problem was there were no racetracks in the South capable of hosting an event of that magnitude. Most of the facilities were small dirt tracks, often a mile or less in length.

Working with fellow farmer Sherman Ramsey, Brasington carved out a track on Ramsey's property in rural Darlington, South Carolina. Ramsey had one request of Brasington—do not disturb the minnow pond on the property.

"I had been talking with Mr. Ramsey for a few weeks about building a track down here," Brasington said. "I told him that I'd already been to Indianapolis and I'd checked into it pretty good, and I was sure it would go. Mr. Ramsey . . . was willing to take a chance. We shook hands on the deal. No contract or no nothing. We started right there—on a handshake."

As a result, Darlington became the largest paved track in the South at that time, with an egg-shaped oval to protect the pond per Ramsey's request.

Once the state-of-the-art facility was completed on May 12, 1950, plans to run a 500-mile race were already in the works.

On December 5 1949, the Central States Racing Association (CSRA) announced it would host a 500-mile event at the brand new Darlington International Raceway, which is now known simply as Darlington Raceway.

The CSRA was not the only sanctioning body planning a 500-mile stock car race for the 1950 season. Bill France's old rival, Sam Nunis and the AAA announced plans to hold a 500-mile race at the one-mile dirt oval Lakewood Speedway in Atlanta, Georgia. While France remained skeptical that stock cars could last a full 500-mile race, Nunis's announcement forced him to act.

Without a location for his planned 500-mile event, France took advantage of the CSRA's low competitor draw. The sanctioning body that primarily hosted races in the Midwest was having trouble attracting drivers for their 500-mile race at Darlington, and France stepped in to help. Always an opportunist, France made a deal with Mason Benner, president of the CSRA, and assured him his drivers would enter the 500-mile race if the race was co-sanctioned by both the CSRA and NASCAR. Once France's deal was done, Nunis abandoned his plans of hosting the 500-mile race at Lakewood Speedway.

With NASCAR now sanctioning the 500-mile race at Brasington's Darlington track, the entry list began to fill up and excitement started to grow. To help attract attention, the purse for the inaugural event would be $22,500, the largest and most lucrative for a stock car event at that time.

A total of 82 cars showed up to make the 75-car field, so qualifying sessions were held over the course of fifteen days. Curtis Turner earned the pole with a lap of 82.034 seconds in a 1950 Oldsmobile. Jimmy Thompson's Lincoln was second, while Gober Sosebee, Bob Flock, Lee Morgan, and Virgil Livengood made up the first three rows. One of the latest entries into the race was former AAA Indy car driver Johnny Mantz, whose Plymouth, co-owned by Bill France and Hubert Westmoreland, was among the slowest in the field throughout the qualifying sessions.

On Labor Day, Monday, September 4, 1950, in front of an estimated crowd of 25,000 fans, the 75-car field took the green flag for the first 500-mile Strictly Stock superspeedway asphalt race.

Georgia native Gober Sosebee led the opening four laps of the 400-lap event which would put the drivers to the test all race long. The dirt track regulars were ill prepared for the high speeds on the high-banked asphalt track, and as a result tires began to wear early and often. Drivers were wearing out tires so fast that crew members began taking tires off passenger cars parked in the infield. Red Byron, who would go on to finish third, went through twenty-four sets of tires during the event.

Curtis Turner and Everett "Cotton" Owens would each take turns leading the field, but they too would be forced to pit for fresh rubber. Despite starting in the 43rd position, Johnny Mantz used his asphalt-racing experience to maintain an average speed of 75.250 seconds to save his tires throughout the race. As others fell by the wayside, Mantz sustained his pace, taking the lead on the 50th lap of the race and never looking back.

Mantz took the checkered flag a full nine laps ahead of second-place Fireball Roberts, with Byron finishing third, Bill Rexford fourth, and Chuck Mahoney fifth. Amazingly, Mantz ran the entire race on the same set of tires. Slow and steady truly did win the race that day.

"Everyone had gone past me," Mantz said, "It wasn't long before I saw cars blowing tires, hitting the pits, and sailing into the guard rails. I kept punching my stop watch and holding a 75 mile-per-hour range."

After that historic day, NASCAR would forever be changed. The cars of the Strictly Stock Series, at the time called Grand

National, had not only survived the grueling 500 miles, but they put on a good show and drew a large crowd. While NASCAR would primarily continue to race on dirt tracks, France knew the future of the sport was on the high-banked superspeedways.

Over the years, the allure and prestige of the Southern 500 at Darlington would grow to enormous heights. Brasington's idea of creating an Indianapolis 500 of the South had done just that. Each year, fans would flock to the egg-shaped track in the middle of rural South Carolina peanut and cotton farms. The race turned into a spectacle with parades, events and infield parties that would rival today's Talladega Boulevard.

The Southern 500 would become one of NASCAR's crown jewel events of which every driver wanted to add his name to the list of winners. Starting with the 1985 season, R.J. Reynolds made it even more lucrative to win the grueling event when they created the Winston Million.

The Winston Million program would reward a driver with a million dollars if he could win three of four premier races on the schedule: the Daytona 500, Winston 500 at Talladega Superspeedway, Coca-Cola 600 at Charlotte Motor Speedway, and the Southern 500. Bill Elliott became the first driver to win the Winston Million in 1985, taking the checkered flag at Daytona, Talladega, and Darlington. The accomplishment thrust Elliott into the national spotlight, landing him the nickname "Million Dollar Bill" and leading to an appearance on the cover of *Sports Illustrated*, the first NASCAR driver to do so.

Despite attempts by Darrell Waltrip (1989), Dale Earnhardt (1990), Harry Gant (1991), Davey Allison (1992), and Dale Jarrett (1996), Jeff Gordon would become the only other driver to earn the Winston Million in 1997.

As NASCAR expanded into the twenty-first century it hoped to race into new markets across the country such as

California, Chicago, Kansas, and others. To expand the schedule and accommodate for additional race dates, NASCAR stripped Darlington Raceway of the traditional Labor Day Weekend race date, granting it to California Speedway starting with the 2004 season. Darlington was limited to one race a year, but the distance remained 500 miles.

On August 26, 2014, NASCAR announced the 500-mile Darlington race would move back to the traditional Labor Day Weekend date starting with the 2015 season.

"These are exciting times at Darlington Raceway. This move was made with the fans in mind because it is what they have been asking for," said then track president Chip Wile. "We look forward to returning to our storied weekend and giving our fans the opportunity to create new Bojangles Southern 500 Labor Day memories in 2015."

The NASCAR Hall of Fame Voting Committee selected Brasington, who sold his stake in the speedway in 1952, for the Landmark Award for Outstanding Contributions to NASCAR in 2015.

"Harold's bold construction of Darlington in 1950 ushered in the superspeedway era and was the catalyst that solidified NASCAR's early future," NASCAR Hall of Fame member Richard Petty said. "Harold was an early pillar of racing and his many contributions and accomplishments deserve to be remembered."

NASCAR Hits the Road

NASCAR MAY have been born on the dirt ovals and high-banked asphalt tracks, but it did not take long for the country's leading stock car circuit to hold a race on a road course turning both left and right.

The eighteenth race of the 1954 season took place on June 13 at a makeshift two-mile road course at Linden Airport in New Jersey. The race was NASCAR's first venture into traditional road course racing. To attract more attention and competitors, Bill France opened the entry list to sports cars as well. It was only the second time France had allowed foreign-made cars in NASCAR competition.

Buck Baker earned the pole position with a lap of 80.536 miles per hour and led the 43-car field to the green flag. Along with the usual Hudson, Plymouth, Oldsmobile, Ford, and Dodge stock cars, the race featured Jaguar, MG, Porsche, and Austin Healey.

Conrad Janis, an actor that would go on to star in the television show *Mork and Mindy*, ran the race under the pseudonym J. Christopher. Janis was an amateur racer and, the pseudonym protected that status. Driving one of the thirteen Jaguars in the field, Janis completed 23 of the 50 laps and finished 39th.

Starting seventh that day was Buffalo, New York, native, Al Keller. Driving a Jaguar owned by big band lead Paul Whiteman, Keller took the lead from Herb Thomas's Hudson 23 laps into the race and never looked back.

Taking the checkered flag on the 50th lap, Keller earned his second NASCAR premier series win in twenty-nine starts and became the first driver to earn a victory in a foreign-made race car. The win would also be Keller's final for Bill France, as he announced after the victory he would leave NASCAR to race with AAA and attempt the Indianapolis 500.

On November 20, 1955, the NASCAR premier series traveled to Lancaster, California, to race on the two-and-a-half mile dirt road course at Willow Springs International Raceway. Jim Reed's Chevrolet led the 37-car field to the green flag in front of a crowd of seventeen thousand people.

Former AAA champion Chuck Stevenson and Marvin Panch swapped the lead seven times over the 80-lap event. Stevenson took the lead from Panch for the final time on the 62nd lap and earned his first NASCAR premier series victory in just his second career start.

The stock cars would make the trip to the 4.1-mile Road America road course in Elkhart Lake, Wisconsin, for the 37th race of the 1956 season. While the race was not officially run in the rain, wet weather caused tricky conditions for the stock car drivers. Tim Flock earned the victory over Billy Myers to win $2,950 for the victory. The NASCAR Xfinity Series began competing at Road America in 2010, while the NASCAR Cup Series returned during the 2021 season.

NASCAR would kick off the 1957 season with two road courses in the first three races of the year. On November 11, 1956, Panch would get his win on the dirt road course at Willow Springs, winning from the pole and earning $1,550 in purse money. On December 30, 1956, the third race of the 1957 season took place on the 1.6-mile paved Titusville-Cocoa Speedway road course in Titusville, Florida and was won by Fireball Roberts.

While the majority of NASCAR premier series races remained on dirt and paved ovals, stock cars would continue to race on road courses throughout the sport's history. The stock car racing circuit stopped at road courses that included Augusta International Raceway in Georgia (November 17, 1963, won by Fireball Roberts), Bridgehampton Raceway in New York (1958, 1963, 1964, and 1966), and Kitsap County Airport in Bremerton, Washington (August 4, 1957, won by Parnelli Jones).

One of the longest active road courses on the NASCAR schedule was Riverside International Raceway in California. The track hosted its first NASCAR premier series event, the "Crown America 500," on June 1, 1958. The 190-lap race featured 46 cars on the 2.631-mile paved road course, including four foreign-made entries—two Citroens, one Goliath, and a Renault. Eddie Gray won that first race over Lloyd Dane in front of a crowd of four thousand.

NASCAR would return to Riverside in 1961, and then return each year from the 1963 season until the 1988 season. From 1970 to 1987, the road course held two races a year, and in 1981 it actually hosted three races. Riverside kicked off the NASCAR premier series in 1965. The road course would again host the season-opening race from 1970 until 1981. In 1982, the Daytona 500 was made the first race of the season, and has remained so ever since.

Rusty Wallace won the final NASCAR race at Riverside International Raceway on June 12, 1988, edging Terry Labonte by just 0.34 seconds.

During the 1990s and early 2000s, NASCAR held road course races exclusively at Watkins Glen International in New York and Sonoma Raceway in California. The premier series first raced at The Glen in 1957, then again in 1964 and 1965 before returning for good in 1986. Three years later the premier series would tackle the hilly terrain of Sonoma's wine country.

In the late 2010s and early 2020s, the NASCAR Cup Series schedule saw an explosion of more road course events than ever before—some at traditional road course facilities, others at purpose-built tracks.

Starting in 2018, the second race of the year at Charlotte Motor Speedway was moved from the traditional 1.5-mile tri-oval to a purpose-built infield road course that also utilized parts of the oval track modified with chicanes. Deemed the "Roval," the new track presented drivers and teams with a host of new challenges never before experienced at Charlotte Motor Speedway. With the event being a cutoff race for the NASCAR Playoffs, the unique course has produced thrilling and memorable action year after year.

Ryan Blaney earned the first win at the Charlotte Roval in dramatic fashion on September 30, 2018, sneaking past Martin Truex Jr. and Jimmie Johnson after Johnson got into the final chicane too hard, slid the car and took out the unsuspecting Truex. Looking at the checkered flag going into the final corner, Truex instead crossed the line in the 14th position. Johnson's last-corner mistake cost him a chance at victory and ultimately led him to miss the cutoff for the next round of the NASCAR Playoffs.

When coronavirus altered the 2020 NASCAR Cup Series schedule, officials were forced to shift their original plans for the season and find other options. The twenty-third race of the 2020 season took place on the Daytona International Speedway Road Course—the traditional home to the annual Rolex 24 Hours of Daytona—on August 16, with Chase Elliott taking the win over Denny Hamlin.

NASCAR returned to the Daytona Road Course layout for the 2021 Busch Clash and the second race of the season. It marked the first time in NASCAR history in which the season started with back-to-back races at Daytona International Speedway utilizing both layouts. Michael McDowell scored his first-career NASCAR Cup Series victory by winning the 2021 Daytona 500, while Christopher Bell earned the win on the road course layout in the O'Reilly Auto Parts 253.

Along with the Daytona Road Course, the 2021 season saw three additional road course events included in the NASCAR Cup Series schedule—Circuit of the Americas, Indianapolis Motor Speedway Road Course, and Road America.

NASCAR president Steve Phelps described the increased number of road courses as something everyone in the sport was ready to see happen.

"I'm thrilled with the schedule. As I said, it's a bold schedule," Phelps told reporters in February 2021. "The fans had said they want more road courses. The OEMs said they want more road courses. Our broadcast partners said they wanted more road courses. As evidenced by what happened on Tuesday night, I think having stockcars on road courses works well. They're slipping and sliding; they get into each other. It puts on great racing."

Circuit of the Americas, a top-notch facility that has hosted Formula One, Moto GP, INDYCAR, IMSA, and other world-class series, hosted its first NASCAR Cup Series event on May 23, 2021. The event was plagued with heavy rain and conditions nearly impossible to compete in. Chase Elliott won the inaugural event, which ended under caution due to the weather conditions. The series returned to COTA in 2022 under much drier conditions, with Ross Chastain, AJ Allmendinger, and Alex Bowman putting on a battle for the ages on the final lap, with Chastain coming out ahead of the field to earn his first NASCAR Cup Series win.

Once a crown jewel event, the annual trip to Indianapolis Motor Speedway received a major overhaul in 2021 by moving away from the traditional oval and to the infield road course that had been previously utilized by Formula One, Moto GP, INDYCAR, and others.

The inaugural race on the layout took place on August 15, 2021, and was marred with issues concerning the curbing around infield sections of the track. As the cars raced through the chicanes and used every inch of the track, the curbing began to come detached from the ground, damaged cars, and caused multiple incidents. NASCAR and track officials were forced to red flag the race for nearly twenty minutes to remove the curbing and make it suitable for racing conditions again.

In another shuffle of the schedule in 2021, the NASCAR Cup Series returned to Road America for the first time since 1956. The 4.1-mile road course not only returned to the NASCAR Cup Series schedule, but it did also so on Fourth of July weekend, a date traditionally reserved for the 400-mile race at Daytona International Speedway, which shifted to the twenty-sixth race of the season and the final event of the regular season.

The sprawling road course provided fans with a perfect camping and relaxed atmosphere, unlike many Fourth of July races before. The outpouring of support from the fans in the Wisconsin area was not lost on Chase Elliott after winning the first race back at Road America on July 4, 2021, calling their passion "amazing" and humbling.

"Just the people in general. Whether you're wearing 9 gear or not, there were just a lot of people here," Elliott said after his victory. "I was just struck by how big of race fans we have and had here today. This place was packed. This is a massive road course. They were literally, I mean, people everywhere around the course.

"It's exciting, man. When they change the schedule up, go to new places, you bring energy and excitement that our series deserves to have. I think we saw that today," he said.

Never an organization to rest on its laurels, NASCAR continued to innovate and look for new opportunities to mix up the schedule and add to their array of road courses. Thanks in large part to iRacing simulation, NASCAR and the city of Chicago planned the first-ever NASCAR street course, taking NASCAR Cup Series cars through the iconic city.

"This is a monumental day in the history of our sport at NASCAR," said NASCAR senior vice president, racing development and strategy, and fourth generation France family member Ben Kennedy. "In seventy-five years our sport has never had a street course, and I am excited and proud to announce to you today that on July 2, 2023, the NASCAR Cup Series will be coming to downtown Chicago and racing here, so really excited about that.

"When you think about Chicago, you think about the rich history of sports here, you think about the iconic skyline, you think about the iconic landmarks around the city, Navy Pier, Cloud Gate, a handful of them, to actually have our NASCAR Cup cars racing around the streets of Chicago is going to be something just incredible," said Kennedy.

"For the first time in our seventy-five-year history, we are going to race a street course," said NASCAR president Steve Phelps. "Not just any city; we're going to race in Chicago. Not at the outskirts of Chicago; we're going to be in downtown Chicago, Lake Michigan, Lake Shore Drive, Michigan Avenue, Columbus. It's not going to be like any NASCAR race ever, not just because it's on a street course, but because of what we are going to do around the development and the hospitality of that racetrack. It will look nothing like any NASCAR race we've ever had."

While stock car racing is traditionally seen as running primarily on dirt and paved ovals, road courses have always played a crucial role in NASCAR history and that role has expanded as the sport looks toward the future. Yet it all started with that first left- and right-turn course at the Linden Airport in 1954.

NASCAR Goes International

LONG CONSIDERED a sport born from the rural American South, NASCAR's earliest years took the stock car circuit up and down the East Coast, and visiting the first venue outside the United States borders did not take long.

Five years after NASCAR was created at the smoky Streamline Hotel in Daytona Beach, Florida, the premier series made the trip to Stamford Park in Niagara Falls, Ontario, Canada, for a 200-lap race around the half-mile dirt track.

The first ever race outside the United States was won by Buddy Shuman, driving a 1952 Hudson Hornet for team owner B. A. Pless. Shuman won by two laps over the field, and only six of the seventeen drivers entered in the event finished the race.

NASCAR returned to Canada on July 31, 1952, when longtime promoter Ed Otto brought the NASCAR late model series to 0.33-mile Canadian National Exposition Speedway. The speedway in Toronto, Canada, hosted late model races for years before the premier series was brought in on July 18, 1958. Over 9,700 fans watched as Lee Petty won over Cotton Owens on the paved short track.

These two events remain the only two premier series points-paying races run outside the United States.

Forty years after NASCAR raced in Toronto, NASCAR headed Down Under for an exhibition race at the newly built Calder Park Thunderdome in Melbourne, Australia. The

1.119-mile paved track featured 24-degree banking in the corner and was built by Bob Jane, who had caught the NASCAR bug after a visit to the United States.

NASCAR's premier series visited the facility and became the first to just use the oval portion of the track when an exhibition race was run on February 28, 1988. The 280-lap event featured 32 cars, with a mix of American NASCAR drivers and Australian natives. It also marked the first time in history NASCAR held a race outside the borders of North America.

Neil Bonnett won the Goodyear NASCAR 500 exhibition race, edging Bobby Allison and Dave Marcis, the only three cars on the lead lap. Other NASCAR stars that took part in the Australian exhibition race included Chad Little, Kyle Petty, Hershel McGriff, and Michael Waltrip.

NASCAR would once again leave North America in the mid-1990s, this time heading to Japan. Under the guidance of Bill France Jr. and Paul Brooks, the sport once again went international for a series of exhibition events.

The 1996 offseason was a busy one for a host of NASCAR drivers and teams as they made the trip to Suzuka City, Japan, for an exhibition race at the 1.394-mile road course Suzuka Circuitland.

The exhibition race was NASCAR's first venture into the Asian market and featured some of the biggest names of the day: Dale Earnhardt, Rusty Wallace, and Jeff Gordon, to name a few.

The stars of NASCAR brought on a host of fanfare from the Japanese, with large crowds jamming the track. The race also featured four Japanese drivers: Keiichi Tsuchiya, Akihiko Nakaya, Kazuteru Wakida, and Hideo Fukuyama.

The 100-lap event featured four lead changes, with the race coming down to a duel between Wallace and Earnhardt. Despite Earnhardt's pressure, Wallace was able to hold on to win NASCAR's first race held in Japan.

The series would return to Japan in 1997 and 1998 for additional exhibition races run after the end of the regular season.

While traveling to Japan for meetings in 1997, Bill France Jr. suffered a heart attack, the result of an arterial blockage. Sent to a former military hospital to have a stent surgically installed, France was stuck in the Japanese hospital due to regulations requiring a patient to remain in the facility for an extended portion of time.

Brooks was stunned to hear the news that France was hospitalized after they arrived.

"It's one thing if you have an emergency when you're in Daytona or Charlotte where you know the community real well," said Brooks, "but when you're somewhere like Japan and you find out your leader is down, well, it's a pretty big shock."

Longtime sponsor Coca-Cola stepped in to help get France out of the hospital and back on track setting up the exhibition races.

Gary Smith, who accompanied France to the hospital, said, "Coke has a huge presence in Japan, and they helped get Bill out of there under the pretense that he was an important American businessman and that the hospital in Narita didn't have adequate enough facilities to enable him to conduct business during his stay, which was true. So we were able to move him to St. Luke's Hospital in Tokyo. That allowed the doctor to save face since Bill was leaving not because of him but because of the hospital."

The 1997 event featured heavy rain on the day of practice and qualifying for the main event. Not to be deterred, NASCAR came prepared with windshield wipers and rain tires. When rain dampened the surface, teams strapped on the rain equipment and hit the track, the first time in the modern era the series had on-track activity in the rain.

The 1998 race would be moved to Twin Ring Motegi in Motegi City, Japan. The race marked the first NASCAR premier series oval race run in Japan. The two previous races were run on the Suzuka road course, but Twin Ring Motegi was a 1.5-mile oval.

Mike Skinner won both races in 1997 and 1998. Although Skinner never scored a points-paying victory in NASCAR's premier series, he holds the record for most NASCAR wins in Japan.

The 1998 event also marked the first time Dale Earnhardt Jr. would have the chance to race against his father, Dale Earnhardt, in a premier series car.

As the sport continued to grow in the 2000s, NASCAR would expand to tracks in both Canada and Mexico for NASCAR's second-tier series, currently known as the Xfinity Series.

From 2005 until 2008, the series hosted races at Autodromo Hermanos Rodriguez in Mexico City, Mexico. Martin Truex Jr. won the first race run at the 2.518-mile road course. That first event featured nine Mexican-national drivers, with Adrian Fernandez finishing 10th. Denny Hamlin, Juan Pablo Montoya, and Kyle Busch would also earn victories at the Mexico City track.

NASCAR made its first return to Canada in 2007 when it visited the 2.71-mile Circuit Gilles Villeneuve in Montreal for a 75-lap Busch (now Xfinity) Series event. Kevin Harvick earned the victory at Circuit Gilles Villeneuve over Canadian Patrick Carpentier to kick off the series' stint north of the border. The series would return to the popular road course until 2012.

In 2008, the Xfinity series made history at the Canadian road course. When wet weather hit the track, rain tires and windshield wipers were put on the cars for the first time in race conditions.

While none of NASCAR's top three national touring series race outside of the United States any longer, the sanctioning body has developed racing series in Mexico (2004), Canada (2007), and Europe (2012).

An Instant Classic: The First Daytona 500

As NASCAR continued to grow, so did the city of Daytona Beach, Florida. Long a home for races on its famous beaches, the city was looking to expand development along the coast in the late 1950s. Faced with a dilemma, Bill France looked at the example set by Harold Brasington in Darlington and saw something even greater.

In February 1954, NASCAR announced the annual Daytona Beach and Road Course race would move to a new Daytona Beach Motor Speedway for the 1955 season. However, due to multiple setbacks, work on the grounds of the proposed two-and-a-half-mile superspeedway would not begin until November 25, 1957.

"My father knew the beach racing couldn't go on forever," said Bill France Jr. "He had long foreseen the day when NASCAR would be a sport with big modern racetracks, and he tried to get the speedway project started earlier than we did.

"He knew that racing on the sand was a novelty," Bill Jr. added. "He also knew that racing on asphalt and concrete was the future. And the way he saw it, what better place to start realizing the future than Daytona Beach?"

The facility, which France renamed Daytona International Speedway before construction, would be something the racing world had never seen before: a massive two-and-a-half-mile paved superspeedway with banking in the corner totaling 31 degrees.

To build the banking in the corners and sweeping tri-oval, contractors dug a man-made lake on the backstretch. France named it Lake Lloyd after J. Saxton Lloyd, who helped complete the paperwork for the speedway's construction.

For most drivers, the high-banked track was awe-inspiring.

"There have been other tracks that separated the men from the boys," driver Jimmy Thompson said of Daytona. "This is the track that will separate the brave from the weak after the boys are gone."

"It's the darndest thing you ever saw," said Buck Baker. "It's the Hollywood of racing. For the man who really wants to race, this is it."

To help lure competitors to the inaugural 500-mile race at Daytona International Speedway, NASCAR offered a purse totaling $67,760. However, drivers would be forced to prove their ability to race on the massive new track by running 25 miles at speeds in excess of 100 miles per hour, a new requirement for NASCAR drivers.

The inaugural "500-mile International Sweepstakes" was open to hardtop and convertible cars alike. France even allowed Jaguars to compete in the strictly stock event at the new facility.

Cars hit Daytona International Speedway's high banks for the first time on February 1, 1959, for a practice session, but had to be closed the next day so that construction crews could complete the guard rail around the outside of the track.

Qualifying for the inaugural 500-mile race at Daytona took place on February 7, with an estimated 6,500 spectators showing up to take in the new spectacle of speed and power.

The first official race held at the two-and-a-half-mile superspeedway was the 100-mile convertible event held on

February 20. Shorty Rollins was able to get by Glen Wood on the final lap of the race to earn the victory over Marvin Panch, and Richard Petty. Wood ended up fourth after taking the white flag in the lead.

A Modified Sportsman race was held on Saturday, February 21, and was won by Banjo Matthews, who lapped the entire field. During that event Junior Johnson, a member of the inaugural NASCAR Hall of Fame class, became the first driver to be disqualified from a Daytona event after officials discovered his fuel tank was too large.

Come race day, Sunday, February 22, Bob Welborn and Shorty Rollins led the 59-car field to the green flag to start the inaugural Daytona 500 and kick off yet another new era for motorsports.

Welborn would lead the first lap from the pole but would go on to battle with "Tiger" Tom Pistone and Joe Weatherly for the top spot over the course of the next 22 laps. Fireball Roberts started 46th in the field but took the lead from Weatherly for the first time on the 23rd lap. Roberts would command control of the race for the next 21 laps until a fuel pump failed, sending him to the pits and ending his day.

Johnny Beauchamp took the lead after Roberts had his issue, but he would have to battle Jack Smith and Pistone for the top spot through the middle section of the 200-lap event. Lee Petty's 1959 Oldsmobile took the lead for the first time with 50 laps to go.

Over the course of those final 50 laps, Petty and Beauchamp would swap the lead a total of ten times, providing a great show for the 41,921 fans watching from the grandstands. As the final laps of the race approached, Beauchamp did all he could to get around Petty. Coming to the checkered flag, the two cars approached the slower car of Joe Weatherly and crossed the start-finish line three-wide in dramatic fashion.

Beauchamp was initially declared the winner of the race and taken to Victory Lane to celebrate. Yet, many others believed Petty had actually beat Beauchamp to the start-finish line and was the true winner of the race. The race results were deemed unofficial, and NASCAR went to work breaking down the evidence to determine the real winner.

Longtime motorsports photographer T. Taylor Warren captured an image of the cars three-wide at the start-finish line but doubt still existed between France and flagman Johnny Bruner.

"I never want to call another one this close," Bruner said. "I want an electronic eye camera, if I have to buy it myself."

Film footage sent by *Hearst Metrotone News of the Week* helped France and NASCAR officials determine Petty was in fact the winner on February 25, after sixty-one hours of investigation.

"The newsreel substantiated that the cars of Petty and Beauchamp did not change positions from the time those other still photographs were taken just before the finish," said France. "Petty is the winner."

Although exhausting and stressful for France, the dramatic photo finish between Petty and Beauchamp made the 500-mile race around the high-banked two-and-a-half mile track an instant classic. Throughout the years, Daytona would become the "World Center of Racing," and the Daytona 500 one of motorsport's biggest events.

"The speedway's true importance, though, has nothing to do with economic impact, race victories, or things like that," said Bill France Jr., who helped his father with construction of the facility. "It has to do with sheer inspiration. It stands as a testament to what people can do if they put their minds to it and their hearts into it."

Speed It Up: NASCAR's First Pit Crew

ONE OF the biggest keys to success in modern NASCAR is to have a lightning-quick pit crew on your side.

As of the 2022 season, Trackhouse Racing's No. 1 pit crew for Ross Chastain was the fastest crew over the course of the entire season with an average four-tire stop time of 11.601 seconds. Modern pit-crews are made up primarily of former professional and collegiate athletes who focus solely on their role as members of the pit crew. They are a crucial part of the team, one that can help win or lose races week-in and week-out. Yet, that was not always the case.

In the earliest days of NASCAR, changing tires and fueling the car were more of a hindrance than an opportunity to gain ground on the competition. Members of the team or friends of the driver would often help with the arduous task of jacking up each side of the car, changing the four tires and fueling the car for the next run. Sometimes, the driver even jumped out of the car to perform the tasks himself.

That all changed when the Wood Brothers from Stuart, Virginia, began showing up to the tracks and competing in NASCAR events. Brothers Glen, Leonard, and Delano Wood were innovative, fast, determined, and always thinking of ways to gain an advantage on and off the track.

"The fact it was taking forty-five seconds to make a pit stop to change two tires and fuel, we got to looking at it and we figured there could be a lot of time saved in the pits," Leonard Wood, a 2013 NASCAR Hall of Fame inductee said. "We looked at it and the more we worked at it the more we gained. You think you're as far as you can go, you keep looking and you still pick up a second you didn't know you had."

The Wood Brothers worked on developing tools that would perform faster than what the competition was used to using: air compression impact wrenches, hydraulic jacks, and so on. Once tools were developed, the team worked on themselves, practicing to the point where everything was a fluid motion, and each move was a natural flow to the next. They never stopped looking for ways to improve, and the results showed on the track.

"You really want to do it where it becomes habit-forming," Leonard said. "When you go to take five lug nuts off, you don't have time to think about going from one to another, you just do it so much it's automatic that your hand just goes on its own. Then you work on your weakest link. You can get your tires on, but you can't get the jack up quick enough, so you go speed the jack up. You get your jack up, tires off, but now you can't get the fuel in. So, you streamline the fuel system so it flows quicker, where they all get it done at the same time."

The Wood Brothers pit stop innovation was so remarkable it even caught the attention of the open-wheel ranks.

In 1965, Ford racing official John Cowley approached Glen at Darlington Raceway to help Jim Clark's efforts in a Colin Chapman–owned Lotus during the Indianapolis 500. The Wood Brothers' reputation for quick stops and innovation was in high demand as Clark hoped to put the Ford-powered

Lotus in Victory Lane in the biggest open-wheel race in the world.

Taking on the task, the Wood Brothers brought the entire pit crew from the No. 21 team as they headed into unchartered territory for a NASCAR organization. Once on site, the Wood Brothers went to work innovating multiple areas of the pit stop, particularly how the fuel flowed into the car. Their tactics were certainly met with skepticism, but their quick and efficient work proved the doubters wrong.

"We went up a week early and we didn't know if these guys were gonna accept us or not because they were a foreign crew," said Leonard. "If they accepted us and welcomed us to be there, it was gonna work. If they didn't, then it wouldn't work as good. But they rolled out the red carpet and acted like they were really glad we were there. They worked with us any way they could, so we just started preparing the car for the race."

All of the training and practice on the NASCAR side of things paid off when it was time to go to work on the biggest open-wheel stage there is.

"I got a little bit nervous," said Delano Wood, who was one of NASCAR's top jackmen. "But when that No. 82 turned off the track onto pit road, I went into 21 car mode. It took the nervousness out of me."

The first stop of the day took only seventeen seconds, stunning many competitors and spectators alike. The second and final stop of the day totaled just under twenty-five seconds, neither time changing tires.

"I think that first pit stop was right around seventeen seconds, so it kind of caught everybody off guard," Leonard recalled. "One of the commentators during the race was Sam Hanks, a prior winner there, and he said, 'You can bet they didn't get it full with a green crew. You can bet they'll be back in.' Well, we didn't come back in. It kind of blew their mind that we made only two planned stops and, I remember, when we finished the last stop, Chapman jumped over the wall and in the middle of pit road yelled, 'Marvelous!' They were thrilled."

Thanks in large part to the quick pit stops, Clark went on to win the race, making the Wood Brothers the first team to play a part in both an Indianapolis 500 and Daytona 500 victory. To further cement their place in the history books, both Clark's victory in the Indianapolis 500 and Tiny Lund's 1963 Daytona 500 victory came without changing tires.

The Wood Brothers' innovations on pit road would serve as a game changer in NASCAR. Whereas before a pit stop was a mandatory hindrance and just part of the race, after the Wood Brothers pit stops would be seen as a way to gain an advantage on pit road.

Despite a rigid and strict rule book, NASCAR never really pushed back on any of the innovations the Wood Brothers were developing on pit road. Instead, they often welcomed them.

"At that time, they weren't trying to stop you from improving your pit stop in any way, other than the fact they didn't want you to have an air tank and an air operated jack with you. That's about the only thing they drew a line on," Leonard said.

The key to their success on pit road was innovation, determination, and most of all, quick reflexes.

"The Wood Brothers, all of us had very quick reflexes. Number one that's what you've got to have," Leonard said. "You can train a guy all day long, all year long, and some guys just aren't as quick as others. So, number one you want to pick a quick man with quick reflexes."

Those quick reflexes and numerous innovations led the Wood Brothers to ninety-nine victories and five Daytona 500 victories. When Trevor Bayne shocked the world in 2011 by winning the Daytona 500 in his first attempt, he gave the Wood Brothers their fifth victory in "The Great American Race."

Ryan Blaney delivered Wood Brothers Racing's 99th NASCAR Cup Series win at Pocono Raceway on June 11, 2017. During Blaney's victory, the No. 21 Wood Brothers Racing pit crew was the fourth-fastest team of the day on pit road, proving over the years the organization remains among the best on pit road.

As NASCAR grew through the 1980s and 1990s, pit stops continued to evolve and speed up. Thanks in large part to the developments of the Wood Brothers, teams up and down pit road were becoming quicker and quicker. By the time NASCAR turned the corner into the twenty-first century, four-tire pit stops were taking less than twenty seconds. In 2015, pit stops were approaching the ten-second mark.

"If you look at how quick they'll change one tire and hit the ground, it's just unbelievable that a human can work that fast," Leonard said in amazement.

Additional changes came to pit road with the introduction of NASCAR's Generation 7 race car, also known as the Next Gen car. An entirely new purpose-built car, the new vehicle also saw one of the most drastic changes for pit crews in the history of the sport. Starting with the 2022 NASCAR Cup Series season, pit crews were forced to adapt to a single lug nut as opposed to the traditional five lug nuts used throughout the majority of NASCAR history.

While the change took teams some adjustment, the speeds showed no signs of slowing. In the 2022 NASCAR Cup Series season finale at Phoenix Raceway, Christopher Bell's No. 20 Joe Gibbs Racing pit crew had a 9.7-second four-tire stop.

For as integral to success as pit crews are in NASCAR today, none of that would be the case without the Wood Brothers from

Virginia. They were the first to modernize the pit stop with their countless improvements and contributions, and both Glen and Leonard Wood are enshrined in the NASCAR Hall of Fame. Glen was a member of the 2012 class, while Leonard went into the Hall as part of the 2013 class.

The Drivers Organize

THERE HAVE been a lot of changes over the course of NASCAR history, but its staunch anti-union stance is one thing that has never wavered. From the earliest days of the sport up through the modern era, it was clear the France family ran the show, and drivers were viewed as independent contractors, not employees.

Bill France Sr. fought off the idea of unions with the threat of closing tracks and banning drivers, often brandishing a pistol just to reinforce NASCAR's official stance.

The sport's first encounter with the idea of drivers organizing a union came in 1960 when Curtis Turner was busy building Charlotte Motor Speedway. During the building process, contractors hit solid granite and large boulders, and ran into a serious dilemma.

With the ground becoming more and more difficult to break through, Turner was losing money and potentially the entire project. Facing uncertainty and in need of a financial investor, Turner looked to the Teamsters Union and its boss Jimmy Hoffa.

"Curtis was overextended, and he turned to the Teamsters Union," said Tim Flock. "He was going to organize the drivers, and the Teamsters were going to give him a lot of money."

Turner, Flock, Glenn "Fireball" Roberts and Buck Baker flew to Detroit, Michigan, to meet with Hoffa and the Teamsters Union.

When the meeting was through, Hoffa and the Teamsters had offered Turner $800,000 to finish building Charlotte Motor

Speedway. In return, Hoffa required that the drivers organize a union, under the umbrella of the Teamsters organization, called the Federation of Professional Athletes (FPA).

Many of the day's drivers had grown frustrated with NASCAR and its dictator-like front man, Bill France. Ever since the initial meeting at the Streamline Hotel in 1947 to organize stock car racers and create NASCAR, France made it very clear he was in charge.

One of the biggest demands focused on the purse money that was paid out by NASCAR at the end of each race. At the time, most race promoters were paying around 7 percent of the actual race purse to the drivers. With the backing of Hoffa and the Teamsters Union, the drivers sought to increase that to 40 percent of the actual purse.

Another of the competitors' concerns was the lack of insurance provided to drivers and their families. Since drivers were considered independent contractors, NASCAR faced no responsibility in ensuring the well-being of a driver or family after their career was over—regardless of how it happened.

When Bobby Myers was killed in an accident with Fonty Flock and Paul Goldsmith in the 1957 Southern 500 at Darlington Raceway, it was the competitors—not NASCAR— that stepped up to support the Myers family.

"When my dad died, he didn't have any insurance, didn't have anything," said Danny "Chocolate" Myers. "I can remember someone bringing up a bucket of change into the house. NASCAR gave us whatever they could collect in a five-gallon bucket."

In addition, the Teamsters Union pushed the drivers to convince NASCAR to allow pari-mutuel betting. Such betting, popular in horse racing, Bill France Sr. vowed to never allow in NASCAR so long as he was in charge.

Once back from their meeting with the Teamsters Union, Turner, Flock, and Roberts went to work. Teamster official Nick Torzeski helped the drivers sell the case for organization and to sign drivers up for their union cards.

"I certainly agreed with a lot of things they were promising, because the sport needed to change and the purses needed to be higher than what they were," NASCAR Hall of Fame member Ned Jarrett said. "Although, I've learned since as a promoter that they were probably paying all they could afford. We didn't know that. As drivers, you look into the grandstands and you see a million people up there and there was probably only ten-thousand."

Eventually, a large number of drivers would agree to join the FPA. Among those that refused to join were Lee and Richard Petty. Lee was bitter rivals with Turner, and did not appreciate his free-living lifestyle either. The movement went on without them, and Turner, Roberts, and Flock were named officers of the newly formed FPA.

With a large group now card-carrying union members, Bill France went to work immediately crushing the idea of an organized group of drivers. France also had the support of track promoters on his side.

"We cannot afford to run under a union," H. Clay Earles, founder of Martinsville Speedway said. "We can't afford to have a race scheduled and have you people strike on us. That's going to put us out of business. And it puts me out of business, it puts you out of business."

France flew to Bowman Gray Stadium in Winston-Salem, North Carolina, where the next race was being held. On August 9, 1961, France held an hour-long meeting with the drivers, banning organizers Turner and Flock.

The man that ran NASCAR with an iron will told the assembled drivers he "won't be dictated by the union" and had a powerful way of backing up the statement.

"I'm going to get a pistol," said France. "I'll tell every one of you drivers who are making your living right now, I'm going to close every track I've got if you stay with this union deal."

None of them did. Every driver that joined Turner and Flock as part of the FPA turned in their union cards and followed the directive of "Big" Bill France.

"I don't think I was really ready to go on board with it because of the union," Jarrett said of leaving the FPA. "Growing up in the South I just didn't know that much about them. You hear things, but you just don't know what would be good and what would be bad. Certainly, though, a lot of the things they were proposing I was in favor of."

To help address some of the concerns drivers had raised in the organization process, France set up a panel of drivers, promoters, and team owners. The Grand National Advisory Board would include two NASCAR officials and was tasked with evaluating the topics drivers wished to discuss.

The 1960 premier series champion Rex White agreed with Jarrett's sentiment but liked the idea of France's new panel better.

"I joined this union. And I've been thinking about it ever since," said White. "Drivers have legitimate beefs. And the drivers want a fair deal and more money. But let's let this board France has appointed decide what's good for racing, not some union. . . . I'll admit the union offer of a retirement plan sold me. But from now on, I'll think for a while before I sign anything else."

Fireball Roberts, who was acting president of the FPA, was one of the final drivers to turn away from the union and return to France's way of doing things in NASCAR.

"My motives in the FPA were quite clear. I simply wanted to better the positions of race drivers, car owners, myself, and racing in general," said Roberts. "I can see now that affiliating the FPA with the Teamsters that we could possibly accomplish more harm than good for racing. I feel if I do anything to hurt the least man in racing, I will be doing a disservice to my fellow drivers who have been my friends for fifteen years. And I'll have no part of it."

Despite the support garnered, France's power was too tough to overcome for Turner, Flock, and the FPA. France was not going to let the two masterminds of the organization effort get off without a hitch.

After the meeting at Bowman Gray Stadium, France banned both Curtis Turner and Tim Flock for life for working with Hoffa and the Teamsters to try and organize NASCAR.

In response, Turner and Flock took NASCAR to court using Hoffa's lawyers, but the trial was held in Daytona Beach, Florida—the home of NASCAR and the France family. With France in his back pocket, the judge hearing the case did little to help Turner and Flock's cause.

"He read comic books all the time in court while our lawyer was trying to explain what we were trying to do, get the 40 percent," Flock said of the judge presiding over the trial. "Bill France owned Daytona, and he owned that old man."

After four years, both Turner and Flock were reinstated by NASCAR, but the message was sent loud and clear. NASCAR and unions would not mix, nor would they ever.

However, eight years later, in 1969, the drivers would once again attempt to organize and place demands on "Big" Bill France and NASCAR.

Much like in 1961, drivers gathered behind closed doors to discuss their desire for retirement, insurance and pension plans. On August 14, 1969, 11 drivers met in Ann Arbor, Michigan and created the Professional Drivers Association (PDA). The newly organized group of drivers would be led by one of the sport's most popular and successful drivers, Richard Petty.

"Our main goals are a retirement plan and insurance plan for drivers, the formation of a uniform pension plan, and driver and crew convenience at the tracks," said Petty. "If we can clean up our sport from the inside out, it will draw more people to the tracks. The promoters will make more money and can undertake the costs necessary to maintain a pension plan."

Petty presided over the PDA as the group's president. Cale Yarborough and Elmo Langley served as vice presidents. The board of directors included Bobby Allison, Buddy Baker, LeeRoy Yarbrough, David Pearson, Pete Hamilton, Charlie Glotzbach, Donnie Allison, and James Hylton.

"We formed an organization because we felt we were foolish in not forming one," Bobby Allison said of the PDA. "Every other major sport has its players' organization. There are definitely things that we have grievances about."

Once again, France stood firm in his defiance of the driver union once he learned of its existence.

"We're not planning to change NASCAR," he said. "We'll post our prize money and they're welcome to run if they want to. If not, that's their business. There are no contracts with NASCAR. But these fellows had better realize that they can't go very far without factory cars. And I'm sure the factories would put someone in their cars if they should think about a strike or something."

Petty promised the PDA would work closely with NASCAR to help improve the sport, but argued there would be no strike.

That all changed when NASCAR made its first trip to the 2.66-mile Talladega Superspeedway in Alabama. Built by France as a bigger and faster version of Daytona International Speedway, Talladega was a high-banked, wide track that produced some of the fastest speeds seen up to that point.

On the opening day of practice on September 9, 1969, Charlie Glotzbach posted the fastest lap of the day at 199.987 miles per hour.

Leading up to the race, drivers were concerned about tire reliability and the fact they would wear out after a two-lap run during the weekend's qualifying sessions. Neither Firestone nor Goodyear had prepared a tire to sustain the nearly 200 mph speeds most drivers were experiencing. Shipments of additional tires were sent to the speedway, but still the drivers—and particularly the PDA—grew deeply concerned.

"I do know we had a real problem," Baker said of the tire situation. What was NASCAR's solution? Slow down.

"It was explained to us that we could back off, run slower," Baker recalled. "In review, maybe that's what we should have done. But I think if you get a bunch of young bootheads like us out on the track, I'm going to tell you, if Cale (Yarborough) passes me, I'm going to pass him back, and pretty soon both of us would have been in the wall.

"Sure, we could have backed off, and we could have run 168 miles an hour until the final three laps and then run 200 (mph) and finished it that way, but that wasn't going to happen," he said, "and as a result most of the drivers boycotted the race, feeling it was going to be too dangerous."

Eventually Firestone would pull their tires out of competition, leaving Goodyear as the lone tire supplier for the inaugural Talladega 500. Firestone's move to withdraw from the race would play a part in the PDA's ultimate decision.

By Saturday morning, Petty and France had multiple confrontations in the garage area, with other members of the PDA joining their second discussion. In the end, France put it simply, "All those who are not going to race, leave the garage area so those who are going to race can work on their cars."

Petty's team was the first to leave the track. In the end, thirty-seven drivers would boycott the inaugural NASCAR premier series race at Talladega Superspeedway.

"Most of us felt the Talladega track is too rough and the tires we have are not safe to race at speeds around 200 mph," Petty said on behalf of the PDA. "It was just that simple. We stick our necks out every time we race. We aren't foolish enough to play Russian roulette. The track is rough and dangerous. We will not race on the track as it is now."

France was determined to go on with the show. The hard-nosed leader of NASCAR filled the field with Grand Touring (GT) cars and the race started as originally planned. On the morning of the race, France issued a statement addressing the boycott.

"I am very much surprised that some of our drivers and car owners would wait until the last day prior to a major race and withdraw their automobiles from a race. Track officials and NASCAR officials worked until the last moment to get the drivers to fulfill their obligations to the fans who traveled from some distance to see the event. Everyone expected they would race.

"It would be unfair to the spectators who traveled to Talladega to see a race to postpone it. It would also be unfair to the drivers and car owners who wish to compete," said France.

And so the show went on. Fans who had purchased tickets to the inaugural race were allowed to exchange it for a ticket to a future race at Talladega or Daytona.

Richard Brickhouse, originally a member of the drivers' organization, announced his resignation from the PDA over the track's public address system the morning of the race and entered the event.

"I joined the PDA at Darlington, but I didn't expect anything like this," said Brickhouse. "I want to race, but I also don't want to make anybody mad. It was a mighty hard decision to make."

Brickhouse would be glad he entered, as the sophomore driver would go on to take the checkered flag and earn the lone victory of his NASCAR premier series career.

With all of the drama between France and the PDA, the motorsports world was certainly talking about what was taking place in Talladega. France changed the entry lists at each track to include a "Good Faith to the Public" pledge that would guarantee any driver or car owner that signed up for a race would actually compete.

The PDA and NASCAR would never have another major run-in, and no subsequent races were ever boycotted again.

There was some good to come out of the tense situation, however.

"The boycott was the best thing in the world to happen to Talladega," said Buddy Baker, "because everybody in the world read about Talladega, and the next race we had there we ran at 200 miles per hour and put on one of the best races ever."

The NASCAR world would go without any other organizing effort for over fifty years, but that all changed on July 7, 2014, when a group of the sport's top team owners announced they had formed the Race Team Alliance.

Led by Rob Kauffman, who at the time was a co-owner of Michael Waltrip Racing, the organized group of team owners hoped to work with NASCAR to lower overall costs, produce better racing, and protect the long-term investment of team owners.

Kauffman did his best to distance the group from being deemed a union.

"With the encouragement of NASCAR and the manufacturers, the teams have met in various forms and forums over the years to explore areas of common interest. This simply formalizes what was an informal group," said Kauffman.

"The key word is 'collaboration.' We all have vested interests in the success and popularity of stock car racing. By working together and speaking with a single voice, it should be a simpler and smoother process to work with current and potential groups involved with the sport. Whether it be looking for industry-wide travel partners or collaborating on technical issues—the idea is to work together to increase revenue, spend more efficiently, and deliver more value to our partners."

However, NASCAR Chairman and CEO Brian France was not thrilled with the organized effort by the ownership group.

"We didn't think it was necessary," France said of the formation of the RTA during an interview with SiriusXM NASCAR Radio on July 21, 2014. "We think the benefits they will arrive at with this association will be much smaller than they do."

The RTA remained quiet and in the shadows, making little news as they worked on their agenda and expanded to include more team owners.

During the 2015 season, NASCAR broke from tradition and approached the competitors, requesting they gather among themselves and vote on representatives to meet with the sanctioning body. While NASCAR had often held large town hall–style meetings with the competitors in the past, this marked the first time the sanctioning body had asked the drivers to unify into one voice to discuss the issues facing the sport.

While NASCAR tried to downplay the significance of the newly formed drivers' council, the competitors were excited about the opportunities and dialogue it created.

"I've got to say I think this is one of the coolest things I've seen happen in this sport since I've been in it," Jeff Gordon, a member of the drivers' council said at Pocono Raceway on June 5, 2015. "I only wish it had happened long before my final year.

I think to have an open line of communication between the drivers and NASCAR.

"I think we are all on the same page and always trying to pursue the best for the sport, but we have done it in a different way; whether individually you go and have those discussions or it's happening via other routes," he said. "To be able to sit in a room and have a lot of drivers that have a unique perspective on it I thought the panel was fantastic. I thought the openness of the conversation was amazing. I thought that it was all very positive."

The drivers' council shied away from calling itself a union, instead pointing out it was an easier avenue to discuss important topics with NASCAR and help improve racing across the board moving forward.

A Test of Endurance: The First World 600

AFTER YEARS of feuding and fighting with Bill France, O. Bruton Smith and the National Stock Car Racing Association (NSCRA) closed shop in 1951. While Smith had constantly rivaled France in terms of hosting races and drawing talent, he was drafted into the US Army in 1950 to fight in the Korean War, leaving the racing business behind for two years. In his absence, France's other longtime rival, Sam Nunis, finally put an end to the NSCRA.

When Smith got back from the war effort, his focus returned to racing. With the NSCRA out of the picture, Smith was forced to work with France to promote NASCAR events. Yet, the young self-made man had a bigger vision.

During his days promoting NSCRA races, Smith was heavily involved in the Charlotte, North Carolina area, and envisioned a racetrack of his own that would rival France's Daytona and Harold Brasington's Darlington. However, Curtis Turner had the same vision. Smith and Turner butted heads over construction of the tracks, with Turner planning a track in Charlotte and Smith looking at land in Concord, North Carolina, less than thirty miles north.

With neither Smith nor Turner having significant financial banking, the two merged forces and decided on a location off

Highway 29 in Concord for the site of their new one-and-a-half-mile speedway. Their plan was bold and grandiose. They envisioned the longest stock car race to date, held on the same day as the world-renowned Indianapolis 500. While the pair first toiled with the prospect of a 501-mile event, they settled on a 600-mile race, the first of its kind. To make the race even more enticing, a race purse of $106,775 was posted.

From the outset of construction in 1959, things did not go as planned. Wet weather left the ground soaked and full of mud. Excavation crews discovered the land was full of large boulders and solid granite. It would push construction costs through the roof and jeopardize the entire project.

"If they'd searched North Carolina for the worst possible place to build a racetrack, that's where they built it," legendary NASCAR mechanic Smokey Yunick once said.

Faced with mounting costs and a potential end to the entire project, Turner and Smith were left searching for ways to pay the construction bills to get the track completed on schedule. The delays and constant search for money to finish the massive construction project forced Turner and Smith to push the date of the race back to June 19, 1960.

The pair's troubles were not over. Contractors constantly sought money for their work, and less than a week before the track was scheduled to open, Owen Flowe ordered his workers to block the final section of track that required paving, demanding the $600,000 owed to him. Turner and a group of strongmen brandished shotguns and Flowe's men backed down. Lights were set up so that workers could work twelve-hour shifts to finish the pavement in time for the race. Work on the pavement would not be complete until the morning of qualifying.

With little time to set, the pavement was a nightmare for the competitors. Patches of asphalt began chunking and coming apart, with drivers left to navigate the less-than-ideal track conditions. Many cars, including Lee Petty's 1960 Plymouth, were outfitted with wire screens to protect the grille and windshield areas.

"Most of the cars looked like army tanks," Petty said. "We knew the track wouldn't hold up for miles with sixty cars on the track. Heck, it wasn't holding up for four laps while one car qualified. We put big screens over the grille and windshield to keep flying rocks and chunks of asphalt out of the radiators. Hopefully, it would keep flying objects out of the driver's compartment. We even put tire flaps over the rear tires on our Plymouths to keep all the debris from flying up into the guy following us."

Petty was not the only driver in the 60-car field who doubted the racing surface would last the scheduled 600 miles.

"The places that have been weak are getting worse. I think we'll be real lucky to get in 300 miles Sunday," said Buck Baker, who qualified fourteenth and would finish fifth in the race.

The night before the race, crews removed eight-hundred tons of asphalt, paved the racing surface in the corners of the track and covered it with two-thousand gallons of liquid rubber sealer.

Despite the concerns and uneasiness about the racing surface, the show went on. When the track opened on the morning of June 19, an estimated 36,000 fans showed up for the inaugural 600 mile event, which Turner and Smith called the "World 600."

Fireball Roberts put his Pontiac on the pole with a lap at 133.904 miles per hour, followed by Jack Smith, Curtis Turner, Cotton Owens, and Ned Jarrett. Roberts led the 60-car field to the green flag and stayed out front for the next 65 laps. The lead would change hands eleven times between Roberts, "Tiger" Tom Pistone, Junior Johnson, Turner, Smith, and Joe Lee Johnson.

When Pistone retired from the race with a mechanical failure, Smith took control of the race, leading the next 193 laps, building a five-lap lead over the competition. However, the troublesome track ruined Smith's chances at victory. While leading the race, a piece of asphalt came loose and punctured the fuel tank on Smith's Pontiac. With fuel pouring out of the car, the team attempted to stuff rags in the punctured tank, and legendary NASCAR mechanic Bud Moore pleaded for a bar of Octagon soap to plug the hole but was unable to do so.

"It was a freak," Moore would later say of the incident. "In stock car racing one never knows what's going to happen. Sometimes it makes a fellow want to throw his hands up and quit."

With Smith out of the running, Joe Lee Johnson assumed the lead for the first time on Lap 353. Johnson's 1960 Chevrolet would lead the final 48 laps to score the victory in the first World 600. Johnny Allen finished second, driving in relief for Johnny Beauchamp, but was four laps behind Johnson.

Bobby Johns, Richard, and Lee Petty rounded out the top five, but NASCAR disqualified the three drivers—along with Junior Johnson, Bob Welborn, Lennie Page, and Paul Lewis—after it was determined they entered pit road illegally. NASCAR had informed the drivers prior to the race to stay off the infield grass area, which was still primarily dirt. Junior Johnson's car spun out early in the race, sliding through the infield and destroying Victory Lane. Petty and the others would also spin through the infield but continued in the race. Days after the event, NASCAR disqualified the seven drivers, stripping them of their finishing order, money earned, and points accumulated.

"It was three, four, maybe five days later when NASCAR told me that I had been disqualified from the 600 for making an improper entrance to the pits," Lee Petty said. "They disqualified the whole bunch of us. Really, what was I supposed to do, having spun out like that?"

Longtime rival and one of the men behind the event, Bruton Smith, was also incensed about NASCAR's decision to penalize the seven drivers days after the event.

"We at the speedway think it is ridiculous to disqualify a man for such a minor thing. We disagree with the NASCAR decision wholeheartedly," he said.

Despite the initial construction and financial woes, terrible track conditions, and controversial ruling by NASCAR, the 600-mile race at the new Charlotte Motor Speedway was ultimately a success. A smaller than expected crowd had witnessed that first race, but the event would grow to become one of NASCAR's crown jewels, along with the Daytona 500 and Southern 500.

The First African American Winner

BASEBALL HAD Jackie Robinson. Basketball had Earl Lloyd. Hollywood had Sidney Poitier.

For NASCAR, it was Wendell Scott.

Born August 29, 1921, in Danville, Virginia, Scott was not the first African American to race in NASCAR's premier division, but he was the first to compete regularly, the first to win a race, and the first to be enshrined in the NASCAR Hall of Fame.

A World War II veteran, Scott honed his skills as a mechanic while serving in the US Army and began his racing career in 1947.

At a time when the United States was deeply segregated, Scott made his mark on the local short tracks in Virginia, all while battling Jim Crow laws and racism along the way.

Right away, Scott was an ace on the local modified circuit, winning over 100 races before ever making his first NASCAR start at the age of thirty-nine.

With no support other than that of his family, Scott made his NASCAR debut on March 4, 1961, at Piedmont Interstate Fairgrounds in Spartanburg, South Carolina. Driving a 1960 Chevrolet No. 87, Scott finished 17th in the race after oil pressure knocked him out of the running after just 52 laps.

Scott would make a total of 495 career starts, but his most historic occurred on December 1, 1963, at Speedway Park in Jacksonville, Florida.

Driving his self-prepared No. 34 Chevrolet, Scott started in 15th position, and worked his way through the field. In a race in which Richard Petty led 103 of the 200 scheduled laps, Scott was able to take control of the lead with 27 laps remaining and put two laps on the entire field.

However, when Scott came to the 199th lap, NASCAR officials did not wave the white flag indicating the final lap of the race. Instead, Scott once again circled the half-mile dirt track, but failed to receive the flag once again. Completing two extra laps, the checkered flag eventually fell on Buck Baker, with the two-time champion going to Victory Lane for the celebration, trophy presentation, and check presentation.

Scott protested the finish, and it was not until later that evening that NASCAR reversed their call, stating there was a scoring error, and declared Scott the winner. The victory, although delayed, marked the first time an African American driver scored a victory in NASCAR's premier division.

"I knew I'd passed Buck ... three times and only made one pit stop for gas and didn't lose a lap," said Scott. "I knew I had won."

Scott took pride in his work and instilled that sense of pride in his sons. Working on his own equipment before, during, and after NASCAR races, Scott was one of the most tenacious and hardest-working drivers in the garage during his time in the sport.

"He was a fierce competitor. Obviously, he was a risk taker. You know, he just loved speed. He loved machinery," his son Frank said on the night of his father's induction into the NASCAR Hall of Fame.

"He said that in order to drive a race car you have to become a part of the race car. You can't be a separate entity. You have to feel it, you have to be a part of it in order to manipulate and drive on those dirt tracks the way he used to do and lean into it and just manipulate.

"Even his peers, you talk to them, they'll talk about what he could do," said Frank. "He could compensate for a lack of horsepower or lack of equipment with his driving ability. That always made him competitive.

"A lot of people have a misconception about competition today compared to some of the competition back then," he said. "Cars were running just as fast back then, and there were teams, automobile manufacturers, that had multiple teams on the track, but we still competed against them. I think that's something that gets lost in the shuffle."

In a sport deeply rooted in the Jim Crow South, Scott faced a litany of boos and jeers nearly every time he showed up to the racetrack. A man of great virtue and character, Scott always taught his sons not to listen to those in the crowd trying to deter their involvement; instead he focused his attention on those that remained silent.

"Daddy had a thing about when we got to the track and we sort of judged the mood of the crowd as we signed into the track, and when we crossed the track, if you can imagine this picture, and the boos and sometimes the boos were worse than boos, but he told us to listen to the fans that we don't hear, and I thought that was so immaculately said because those were the fans that really supported us," said Wendell Scott Jr.

"Over the years there came to be more of them than we realized. There were more fans supporting us probably during some of the worst periods or epochs of our history of this country. The boos weren't really that much, proportionately speaking."

Despite the adversity he faced both as a person of color and an independent driver and team owner, Wendell Scott never let the pressure of the adversity get the best of him.

"Daddy would settle a matter," he said. "Don't think that he'd just roll over. He would handle it in a gentlemanly way if you would, whatever you call the gentlemanly way. He would put a stop to it if it got to be threatening to either his family or our livelihood."

Despite the rampant racism and obstacles set before Wendell Scott, a number of NASCAR's most prominent drivers of the day found ways to help out.

NASCAR Hall of Fame members Richard Petty, Ned Jarrett, Glen and Leonard Wood, and Rex White were among those in the garage that admired the tenacity and determination of Scott, and were willing to help in any way they could.

"If you accomplish something, people think you do it by yourself. But that's not how it is," said Frank Scott. "Somebody helps you. Somebody helped them, and they pass the blessings on. I remember the first high-rise intake we got, we got from the Wood Brothers, but they went to Tunnel Port. So they passed on some things to us and we were able to do good and compete with it.

"Rex White, we'd go down to his garage. He had a factory deal and had no problems with parts, you know. They had to keep their cars looking spit-shine. They'd take a fender off, or a door, I'd load on the back of the truck, straighten it out, and put it on our car. Ned was the same way."

That help and admiration went far beyond the track as well.

After suffering career-ending injuries in a massive wreck at Talladega Superspeedway in 1973, Scott lost the livelihood he had supported his family with for over a decade, racing just once more after the incident.

One of NASCAR's biggest names, Richard Petty, stepped in to help the Scott family during its time of need.

"He probably wouldn't want me to say this, and maybe I shouldn't. But when my father got hurt at Talladega, he helped give my father support to take care of my family," said Frank Scott of Petty. "He did more than NASCAR. A whole lot more."

While Scott was the first African American to win in NASCAR, he was not the first to compete in the sport. Two other African American racers also made starts in NASCAR's premier division before Scott's first start in 1961.

On July 31, 1955, at Bay Meadows Speedway in San Mateo, California, a local business owner name Elias Bowie entered a No. 60 1953 Cadillac into the 34-car field. Starting in the 31st spot, Bowie completed 172 of the 252 laps and finished 28th, earning $90 for his efforts.

One year later, Charlie Scott became the second African American to compete in a NASCAR race when he drove a Karl Kiekhaefer–owned Chrysler 300 to a 19th-place finish out of the 80 cars entered in the event.

As with Bowie, the 1956 Daytona Beach and Road Course race was Charlie Scott's lone NASCAR start.

Mired in obscurity and the ghosts of NASCAR's past, Elias Bowie and Charlie Scott—perhaps unknowingly—broke the mold and set a course for others to follow in their footsteps. It would take another five years for Wendell Scott to make his NASCAR debut and truly open the doors for minority competitors in NASCAR.

However, when Wendell Scott made his venture into NASCAR's premier division, he did so for the long haul. Unlike those that came before him, Scott was dedicated to the sport.

"There wasn't anyone else on his tier," said Frank. "I know Charlie Scott, he was a good friend of ours, I think he drove for Kiekhaefer. It's apple and oranges in comparison. Charlie's a great friend. We lived with him, he lived with us."

Once Scott stepped away from the sport, it took some time for another African American to make it into NASCAR.

Willy T. Ribbs competed in three NASCAR Cup Series races in 1986 but struggled mightily. The open-wheel racing regular returned to run the full Camping World Truck Series schedule in 2001, but his struggles continued.

In 1999, Bill Lester became the first African American driver to start a NASCAR Busch (now Xfinity) Series race, finishing 21st at Watkins Glen International. Lester was also the first African American to compete in the Craftsman Truck Series, making his first start at Portland International Raceway and finishing 24th.

Lester would go on to qualify for two NASCAR Cup Series races in 2006 driving for team owner Bill Davis. In those two events, Lester finished 38th at Atlanta Motor Speedway and 32nd at Michigan International Speedway.

In today's NASCAR, the doors of opportunity are greater for African American drivers thanks to the NASCAR Drive for Diversity program, Max Siegel's Rev Racing, and others.

The most successful of that new, young crop of minority drivers is Darrell "Bubba" Wallace Jr. Right from the start, it was clear Bubba was going to make his mark on the sport.

In his first NASCAR K&N Pro Series East start on March 27, 2010, at Greenville-Pickens Speedway in South Carolina, Wallace led 22 of the 156 laps and battled with Cole Whitt to score his very first NASCAR win in his very first start.

A graduate of the Drive for Diversity program, Wallace worked his way through NASCAR's home tracks ranks, had a successful run with Kyle Busch Motorsports and Joe Gibbs Racing in the Camping World (now Craftsman) Truck and Xfinity Series, and in 2015 landed a full-time Xfinity Series ride for Roush Fenway Racing.

On October 26, 2013, Wallace led 96 of 200 laps at Martinsville Speedway to earn his first-career Truck Series victory. By earning the win, Wallace became the first African American to earn a victory in a NASCAR national touring series race since Scott's victory in 1963.

Wallace Jr. would go on to score four more Camping World Truck Series victories during the 2014 season, including a race at Martinsville Speedway in which he drove Scott's familiar No. 34 and carried a paint scheme honoring Scott's induction into the NASCAR Hall of Fame.

"I know I had a guardian angel looking over me this weekend," Wallace said after his 2014 Martinsville victory. "To be able to put it in victory lane, you couldn't ask for a better weekend. You thought last year was special, but this definitely beats it."

Wallace was not done making history.

On June 11, 2017, Wallace competed in his first NASCAR Cup Series race at Pocono Raceway driving the iconic No. 43 for Richard Petty Motorsports. Wallace was filling in for an injured Aric Almirola. Teaming up with "The King," Wallace started 16th and finished 26th in his NASCAR Cup Series debut.

He would run four races for the team before Almirola returned from injury, but for the 2018 season Wallace was tapped as the full-time driver of the No. 43 Richard Petty Motorsports Chevrolet.

In the season-opening Daytona 500 that year, Wallace edged out Denny Hamlin at the checkered flag to finish second to Richard Childress Racing's Austin Dillon. The runner-up finish was the highest finish of an African American driver since Scott's win in 1963.

While Wallace was celebrating his second-place finish, another African American was celebrating another historic first. With Dillon returning the No. 3 RCR Chevrolet to Victory Lane in the Daytona 500 for the first time since Dale Earnhardt accomplished the feat twenty years earlier, jackman Derrell Edwards became the first African America pit crew member to win the Daytona 500.

For Wallace, the strong runs would continue but he would also deal with struggles on and off the track. During the 2020 season, Wallace was thrust to the center of attention as the nation struggled with racial injustices, the murders of multiple unarmed African Americans, COVID-19 restrictions, and more.

Seeing a video online of two white men in Georgia shooting an unarmed African American named Ahmaud Arbery, Wallace was moved to the core and disturbed like never before.

"I clicked on it, and I watched it, and I was like, 'What in the hell? What just happened?' I watched it again and I was just dumbfounded at what I had just witnessed," Wallace said. "A black man chased down by civilians and shot and killed."

With rising tensions in the country due to COVID-19, the murder of George Floyd finally moved Wallace to action. On May 27, 2020, Wallace posted to Twitter saying, "Shits getting old.. hell its been old. Wtf is gonna change?!" with a praying hands emoji and the hashtag #GeorgeFloyd.

In an interview with "NASCAR Race Hub" on FOX Sports 1, Wallace read a text he received from his mother following Floyd's death. He told the panel of hosts including Kaitlyn Vincie, Ricky Craven, and Drew Blickensderfer, "My mom said, 'Good morning. I pray as the mom of a black son that I never have to hear you crying out, "I can't breathe." I love you, Bubba, and your life matters to me.'"

Wallace and each of the hosts were left speechless as they tried to wrangle their emotions. Nothing like this had truly ever been openly discussed on a NASCAR program before. The sport that was created and rooted in the South and at one point openly supported and flaunted the Confederate Flag, was now faced with racial tensions that extended beyond the sport.

On June 7, 2020, Wallace wore a T-shirt that read, "I Can't Breathe | Black Lives Matter," and a mask with an American flag during pre-race festivities at Atlanta Motor Speedway. Seeing Wallace's shirt on pit road, African American NASCAR official Kirk Price took a knee during the National Anthem.

"To see Bubba take that stand, and people marching in the streets, I had to do something. Taking a knee was just a small gesture," said Kirk. "A lot of people thought me taking a knee was radical. I don't see it as radical. Now, what Mr. Wallace Jr. did, that's radical."

The incident immediately thrust Wallace into the national news spotlight, with cable TV from both sides of the political aisle weighing in with their opinions on the matter.

"I have a responsibility," Wallace said. "It's automatic that I have a responsibility because of my background, my heritage, my family."

In an interview with CNN on June 8, 2020, Wallace took it one step further and called on NASCAR to ban the Confederate flag at all tracks.

"No one should feel uncomfortable when they come to a NASCAR race," he told host Don Lemon. "So, it starts with the confederate flags, get them out of here. They have no place for them."

NASCAR had requested fans to not fly the Confederate flag at their events in 2015, even offering to exchange the Confederate flag for a new American flag, but it was not an outright ban.

The backlash from some within the NASCAR community and more from the wider public was harsh, hurtful, and at times very threatening. Through it all, Wallace maintained his composure and tried to do his job behind the wheel.

On June 10, 2020, NASCAR rolled into the oldest track on the schedule, Martinsville Speedway, just over thirty minutes from Wendell Scott's hometown of Danville and announced they would officially ban the Confederate flag from all their events for the first time in the sport's history.

"The presence of the Confederate flag at NASCAR events runs contrary to our commitment to providing a welcoming and inclusive environment for all fans, our competitors and our industry. Bringing people together around a love for racing and the community that it creates is what makes our fans and sport special. The display of the Confederate flag will be prohibited from all NASCAR events and properties," the statement read.

That same day, Wallace's No. 43 Chevrolet carried a special scheme with #BlackLivesMatter and a peace symbol with hands of different colors inside on each quarter panel, while the hood featured a pair of hands clenched together in unity—one black, one white—and the words, "Compassion, Love, Understanding." Wallace again wore his "I Can't Breathe | Black Lives Matter" T-shirt and American flag mask during pre-race activities.

Six days after banning the Confederate flag, NASCAR appointed Brandon Thompson to the newly created role of Vice President, Diversity and Inclusion. A longtime industry insider and one of the most prominent African Americans in the sport's leadership, Thompson was asked to further diversify and grow NASCAR to new reaches and to communities traditionally not touched by the sport. He also took over the NASCAR Drive for Diversity Program and led the sanctioning body's alliance with social justice and inclusion organizations.

"NASCAR has made significant progress in the areas of diversity and inclusion, and we now have an opportunity to build on our momentum—both as a company and industry," Thompson said in a statement. "It is with great passion and energy that we will champion our sport as accepting and welcoming of all individuals interested in being part of the NASCAR family."

Three weeks after Wallace's display at Atlanta Motor Speedway, NASCAR headed to Talladega Superspeedway in Alabama, the heart of the deep South. With COVID-19 slowing and restrictions easing, this marked the first weekend since the pandemic broke out in which limited fans were allowed to attend the race. Throughout the weekend, a plane flew above the racetrack carrying a banner with a Confederate flag and the words "Defund NASCAR."

On Sunday, June 21, 2020, a rope tied in a noose was found in Wallace's garage stall by a crew member on his No. 43 Richard Petty Motorsports team. NASCAR security was notified, then senior management of the sanctioning body, who then decided the following day to notify the Federal Bureau of Investigations to involve them.

"As we have stated unequivocally, there is no place for racism in NASCAR. This act only strengthens our resolve to make this sport open and welcoming to all," NASCAR President Steve Phelps told the assembled media on June 22, 2022.

Phelps went on to say that if it was determined that anyone in the NASCAR community did this, they would "unequivocally" be banned from the sport for life.

The NASCAR community immediately came to Wallace's side. A stencil spelling out "I stand with Bubba" was added to the tri-oval grass, and prior to the race his fellow competitors did something unexpected and truly moving.

Wallace's No. 43 Chevrolet was slated to start in the 24th position in the GEICO 500, but while he was sitting in his car with headphones on his fellow competitors approached the car and began to push it to the front of the grid. In a show of complete solidarity, each of the drivers and most of the crew members on pit road walked behind Wallace's car and stood behind him as he emerged from his car.

Clearly emotional from the entire weekend, he folded his arms and buried his head on the roof, as his team owner and NASCAR legend Richard Petty came up and comforted his driver. The entire field stood in solidarity with Wallace as the national anthem was played before the race.

Following an investigation by the Birmingham branch of the FBI, it was determined the rope was a garage pull used to shut the roller doors and had been in place the year before Wallace was assigned the garage stall. Following the ruling, Phelps said NASCAR's goal after the rope was brought to their attention was look out for Wallace's well-being. He also indicated the sport did a complete sweep of their tracks and garage stalls at all facilities, finding only eleven in total with a pull-down rope tied in a knot and "only one noose: The one discovered on Sunday in Bubba Wallace's garage."

"Upon learning of and seeing the noose, our initial reaction was to protect our driver," he said. "We're living in a highly charged and emotional time. What we saw was a symbol of hate and was only present in one area of the garage, that of the 43 car of Bubba Wallace. In hindsight, we should have, I should have used the word 'alleged' in our statement."

Wallace made another appearance on CNN on June 24, addressing NASCAR's decision to ban the Confederate flag following his first appearance on the show, the incident that took place at Talladega, as well as the FBI's findings.

"Don, the image that I have and that I have seen of what was hanging in my garage is not a garage pull," Wallace said. "I've been racing all of my life. We've raced out of hundreds of garages that never had garage pulls like that. So, the people that want to call it a garage pull and put out old videos and photos of knots as their evidence, go ahead. The evidence we have, that I have, it's a straight-up noose. The FBI called it a noose. NASCAR leadership has stated it a noose. I can confirm that. I actually got evidence hanging in my garage, over my car, around my pit crew guys, that it was a noose. I've never seen anything like it."

The incident thrust NASCAR and in particular, Bubba Wallace, into the national spotlight, and again, the NASCAR community came to Wallace's side and supported the driver through and through.

Throughout it all, Wallace served as an ambassador for his sport, his family, and his heritage. He never let his detractors get the best of him, and he took the high road throughout serving as a role model, emulating the example put forth by Wendell Scott so many years before.

Later that year, Wallace was selected to drive for a new NASCAR Cup Series team co-owned by fellow racer Denny Hamlin and NBA legend Michael Jordan. A lifelong racing fan because of his father, Jordan grew up going to races and was now taking that passion to the next level by owning a NASCAR team, which would be called 23XI Racing.

While the expectations were high for the team and Jordan demanded excellence, Wallace and the new organization struggled to get going throughout the early stages of the 2021 season. The team did not score its first top-ten finish until a fifth-place finish at Pocono Raceway. When the season rolled around to Daytona in August for the final race of the regular season, Wallace still had an opportunity to make the NASCAR Playoffs if he could win. Wallace led eight laps and was in contention all night but finished second to Team Penske's Ryan Blaney.

On the first weekend of October, the NASCAR Cup Series returned to Talladega Superspeedway. The site of so much controversy one year prior, Wallace would again become the center of attention that weekend.

The YellaWood 500 was plagued by rain, forcing the race to be postponed from Sunday to Monday, October 4, 2021. While the weather was better than Sunday, the threat of heavy rain persisted the following day and few believed the race would make the final distance of 188 laps. With rain on the horizon, the intensity level in the pack ramped up even more than usual as the field fought for positions, not knowing when a caution might end the race.

Wallace and longtime spotter Freddie Kraft worked wonderfully together as the No. 23 Toyota raced to the front of the field and blocked to keep the hungry pack at bay. On Lap 177, the rain came, and the caution was thrown with Wallace in the lead ahead of Team Penske drivers Brad Keselowski and Joey Logano.

After a red flag delay in which the rain became heavier, NASCAR called the race official and declared Wallace and his No. 23XI Racing team the winners of the event. This marked the first time since Wendell Scott's victory in 1963 in which an African America driver won a NASCAR Cup Series race.

For all the hate he received, all the threats of violence, the racists comments, Wallace remained committed to becoming a winner in the NASCAR Cup Series and had finally accomplished that feat.

"I never think about those things. When you say it like that, it obviously brings a lot of emotion, a lot of joy to my family, fans, friends," Wallace told the TV coverage once the race was declared official. "It's pretty damn cool, just got to be a winner in the Cup Series."

Although some were quick to point out the race was rain-shortened, they had nothing to say the following year when on September 11, 2022, at Kansas Speedway, Wallace scored his second career NASCAR Cup Series victory by leading 58 of the 267 laps and taking the checkered flag one second ahead of team owner Denny Hamlin.

Wallace is far from the only African America to find success in the sport, either. As mentioned earlier, Derrell Edwards became the first African American pit crew member to win the Daytona 500 when Austin Dillon scored the victory in 2018.

A former basketball player at High Point University in North Carolina, Edwards made his transition to NASCAR with RCR after his 2014 season. Four years later he was celebrating in Victory Lane in the sport's biggest race.

Edwards was debating a career playing professional basketball overseas when he was presented the opportunity to join NASCAR as a pit crew member. He recalled a conversation with his father-in-law in which he told Edwards to pursue a career in NASCAR and "become the Jackie Robinson" of pit crews.

"I didn't get in the sport to make history or anything like that," Edwards said. "Honestly, where I come from, I just wanted to show kids you could do other things. My culture and where I come from, you're told you can play basketball or football to be a professional athlete or pursue rap, and it's kind of like a box that they put you in. For me taking on NASCAR, I wanted to open up another avenue or lane to show kids you can do this too and to not be afraid to do other things."

Coming from a basketball background, Edwards admits he did not know much about the history of Wendell Scott or the other African Americans that came before him in the sport. He admits he was more focused on those making a difference now, people such as Bubba Wallace, Dion Williams, and others.

"I learned a lot. I didn't know the magnitude of it," he said. "It's a super cool story, you know. It takes guys like (Wendell), it takes guys like Bubba, next will probably be Rajah Caruth, it takes those people to take leaps to make a difference."

Having found success, Edwards believes his impact is starting to make a difference in creating new NASCAR fans from the African American community. After his Daytona 500 win, people from his community started following the sport more closely, watching races on a weekly basis, and learning more about the sport.

"The culture definitely shifted, and people were more intrigued and interested in the sport," he said. "More people started diving into the sport. Honestly, the whole culture of our sport started innovating and taking steps in the right direction since I've been in it.

"People were just intrigued and didn't know the magnitude of NASCAR, how big it was, didn't know it was on TV every week," he added. "People definitely started asking questions."

While Edwards was not the first African American to serve on a pit crew, he was the first African American to work as a jackman on the iconic No. 3 Richard Childress Racing Chevrolet. He recently signed a five-year contract extension with Joe Gibbs Racing, the first of its magnitude and length for any pit crew member.

Perhaps Edwards would have never received his opportunity without Dion Williams. A student athlete at Wake Forrest University, Williams first joined Chip Ganassi Racing in 2005 and eventually became the assistant pit crew coordinator. After a stint at Richard Petty Motorsports, Williams found a home at Hendrick Motorsports, where he became one of the top pit crew members in the sport and was featured on the TV show *Undercover Boss* with NASCAR's Steve Phelps.

Throughout his career, Williams scored sixteen NASCAR Cup Series victories and was the first African American pit crew member and race shop mechanic for Hendrick Motorsports. After retiring from pitting cars, "Rocko" (as he is known) continued to make a lasting impact on the sport. He works with the NASCAR Drive for Diversity program to help bring minority athletes into NASCAR and help them along the way to success.

One of those athletes is Brehanna Daniels. A graduate of the NASCAR Drive for Diversity Program, Daniels became the first African American female to serve on a NASCAR pit crew when she went over the wall to change tires in 2017.

The increased diversity is not only happening on the track and on pit road, either.

In August 2020, Erik Moses was named president of Nashville Superspeedway ahead of NASCAR's return to the 1.333-mile track. In doing so, Moses became the first African American track president in NASCAR history. In 2022, Latasha Causey became the first female African American to hold the position of track president when she was appointed the position at Phoenix Raceway.

The 2023 Daytona 500 also saw history made when Ricky Stenhouse Jr. took the checkered flag for JTG-Daugherty Racing, owned by Tad and Jodi Geschickter and former NBA star Brad Daugherty. This marked the first time a female co-owner and an African American co-owner scored a victory in the Daytona 500.

"We've got a lot of diversity on our race team, throughout the garage, and it's cool to have two on our race team and put them in Victory Lane here at the Daytona 500. It's super special, and NASCAR is leading the way in a big way," Stenhouse said following the race. "It's cool to play a small part of getting them to Victory Lane."

There is no doubt African Americans and other minorities still face an uphill battle to break into NASCAR's highest ranks, but the example set forth by Wendell Scott and carried on by Bubba Wallace has given each driver, crew member, team owner, and industry insider following in their footsteps something to strive for. Over the past handful of years, NASCAR has taken great strides to make the sport more welcoming and open to all people from all backgrounds, but the work is never done.

The First Ladies of NASCAR

WHILE NASCAR history has been dominated by men, the women of the sport have never been far behind—both on and off the track.

In recent years, Danica Patrick served as the face of women in the sport, making waves as a full-time driver in the NASCAR Cup Series. The popular, good-looking, hard-driving racer transitioned from open-wheel racing to stock cars in 2010.

With sponsor support from GoDaddy and Dale Earnhardt Jr.'s JR Motorsports, Patrick made her first NASCAR Nationwide (now Xfinity) Series start on February 13, 2010 at Daytona International Speedway. Working with crew chief Tony Eury Jr., Patrick competed in twenty-five NASCAR Xfinity events from 2010 to 2011, all while running the full seventeen-race IndyCar season.

Patrick made her first NASCAR premier series start in the 2012 Daytona 500, finishing 38th. Driving for team owner Tommy Baldwin Jr., Patrick competed in ten NASCAR Cup Series events that season, finishing a season-best 17th at Phoenix Raceway. That same year, Patrick made history as the first female to qualify on the pole for the season–opening NASCAR Xfinity Series race at Daytona.

By making her first start in the Daytona 500, Patrick became the second female driver to race in both the Indianapolis 500 and Daytona 500, joining racing icon Janet Guthrie. A legend in her own right, Guthrie was the first woman to compete in the World (now Coca-Cola) 600 at Charlotte Motor Speedway after receiving an invite from track promoter H. A. "Humpy" Wheeler in 1976.

Attempting to qualify for her first Indianapolis 500, Guthrie was one of the biggest storylines in the month of May that year but faced backlash from many in the open-wheel world. For Wheeler, who was in his first year serving as promoter at Charlotte, Guthrie and her adventures were taking attention away from his World 600. With the competition fierce, Guthrie struggled for speed in Indianapolis. Wheeler went to work doing his best to lure the news sensation to his track and steal the glory and attention back from Indy.

"As [the race] got closer, I called her every day. Finally, she said, 'If I don't make it up here, I want the best possible car I can get to run Charlotte,'" said Wheeler.

When Guthrie failed to qualify for the Indianapolis 500, her attention—and that of much of the motorsports world—turned to Charlotte and the World 600.

Driving a No. 68 Chevrolet owned by Lynda Ferreri, Guthrie struggled to adjust to the stock car after a month of racing open wheel cars at Indianapolis. Seen as an outsider in the NASCAR garage as well, one of the sport's biggest figures stepped forward to provide some direction and get her on track.

Team owner Junior Johnson told his driver Cale Yarborough to drive the No. 68 Chevrolet for a number of laps in practice. Yarborough, who would go on to win the 1976 NASCAR Cup Series championship, could only better Guthrie's lap times by a little bit.

"Junior looked at me and said to his chief mechanic, 'Give them the setup.' I finally had a drivable car on my hand," said Guthrie.

In a race dominated by David Pearson, Richard Petty, and Yarborough, Guthrie held her own and finished 15th out of 40 cars, completing 379 of the 400 laps. In her first NASCAR Cup Series start she was able to finish ahead of drivers such as Richard Childress, Bill Elliott, Buddy Baker, Dale Earnhardt, Bobby Isaac, and more.

"There is a lot of pride in what she did and the guys who worked on the car. She did a lot more than anyone thought she would do. I knew she could drive the race car the first time she got in it," said Ralph Moody, Guthrie's crew chief that day.

Even race-winner Pearson was impressed with what he saw out of Guthrie's first efforts in one of NASCAR's most grueling events.

"She wasn't really a problem. I lost a couple of seconds to Richard (Petty) as I tried to go outside and she went outside, then I started under and she came down. She was trying to pass someone else, too, just like I was," Pearson said after the race. "I felt like she wouldn't last 600 miles. She did surprise me and really surprised a lot of people."

Guthrie competed in four more events during the 1976 season, and on February 20, 1977, she became the first female driver to compete in the Daytona 500. Running her now-familiar Kelly Girl-sponsored No. 68 Chevrolet, Guthrie started 39th on the field and was still running at the end of the race, finished 12th, and completed 188 of the 200 laps. She was the highest-finishing rookie driver in the field. During the 1977 season she qualified ninth and finished sixth at Bristol Motor Speedway, a mark that would last until Patrick broke onto the scene nearly three decades later.

On February 17, 2013, Patrick made history by becoming the first female to qualify on the pole for a NASCAR premier series race. Her qualifying run of 196.434 miles per hour earned her the pole for the 55th running of the Daytona 500.

By besting the other forty-five cars in the field, Patrick accomplished something never before done in the 2,354 previous NASCAR races. Like a true racer, she credited her Tony Gibson–led crew for making her Stewart-Haas Racing Chevrolet fast and as easy to drive to the top of the charts.

"I feel like, first and foremost, I grew up with good values and good goals. I was brought up to be the fastest driver, not the fastest girl," Patrick said after that historic qualifying run in 2013. "That was instilled in me from very young, from the beginning. Then I feel like thriving in those moments where the pressure's on has also been a help for me. I also feel like I've been lucky in my career to be with good teams and have good people around me. I don't think any of it would have been possible without that.

"For those reasons, I've been lucky enough to make history, be the first woman to do many things," she continued. "I really just hope that I don't stop doing that. We have a lot more history to make. We are excited to do it."

Unlike those that came before her, Patrick has used her heavy right foot, marketability, and approachable personality to open the door for new fans and serve as an example to young girls watching her race.

"It's also nice to hear families talk about the fact that a little girl might say, 'But, mommy, daddy, that's a girl out there.' Then they can have the conversation with their kid about you can do anything you want and being different doesn't by any means not allow you to follow your dreams," she said at Daytona International Speedway in 2013. "I love to think that conversation happens in households because of something I'm doing."

During that same Dayton 500, Patrick became the first female in NASCAR history to lead a premier series race, pacing the field for five laps in the 'Great American Race.'

However, the Daytona 500 pole and laps led were not Patrick's first, first. Before making her way to NASCAR, Patrick set a number of high marks in the IndyCar Series as well.

In her first Indianapolis 500 in 2005, Patrick started fourth and finished fourth driving for team owners Bobby Rahal and television late-night host David Letterman.

During the 2009 running of the Indianapolis 500, this time running for Andretti Green Racing, she finished third behind race-winner Helio Castroneves and Dan Wheldon. With that third-place finish, history was made yet again as Patrick became the highest-finishing female racer in the Indianapolis 500.

On April 20, 2008, Patrick became the first woman to win an IndyCar Series race, also driving for Andretti Green Racing. The victory came in as she passed Castroneves with three laps to go at Twin Ring Motegi Superspeedway in Motegi City, Japan. The historic win occurred in her 50th IndyCar start.

"Finally! This was a long time coming," she said after the victory. "It was a fuel strategy race, and my team called it perfectly for me. Even before they started to tell me to save fuel, I was doing it. It's a difficult balancing act when you're trying to make a specific fuel mileage 'number' without losing positions on the track. I knew I was on the same [pit] strategy as Helio and when I passed him for the lead, I couldn't believe it. This is fabulous! To get my first win is great. To do it at Honda's track is even better."

Patrick's accomplishments and firsts continued once she made the transition to NASCAR's stock car racing.

She currently holds the record for the highest-finishing female in a NASCAR touring series race, fourth in a Nationwide (now Xfinity) Series 300-mile event at Las Vegas Motor Speedway on March 5, 2011.

She also currently holds the record for the highest finishing female in NASCAR national touring series championship year-end standings. Driving for JR Motorsports, she finished 10th in overall points in 2012.

Throughout her seven-year NASCAR Cup Series career, Patrick completed in 191 races with seven top-ten finishes, one pole, and 64 laps led. Patrick's final NASCAR Cup Series start took place on February 18, 2018, in the 60th running of the Daytona 500, where she finished 35th after being collected in a crash just past the 100-lap mark.

While Danica Patrick set a path for female racers in NASCAR's top division and accomplished many firsts, she was certainly not the first woman to compete in a NASCAR stock car race.

The involvement of women in the growth and success of NASCAR has always been prominent, whether behind the scenes or behind the wheel. And their involvement goes back to the sport's very beginnings.

As 'Big Bill' France was ruling NASCAR's earliest days with an iron fist, a flexible rule book, and the occasional pistol, his wife Anne Bledsoe served as secretary for the new sport.

"[She was] a great ambassador to our sport and someone who always strived to grow the business," Lesa France Kennedy said during the 2015 NASCAR Hall of Fame Induction Ceremony, where 'Annie B,' as she was known, was awarded the Hall's Landmark Award for Outstanding Contributions to NASCAR. "She did it her way in an honest, often understated and pragmatic fashion. She was the glue that held NASCAR together in the beginning, and she worked tirelessly to see it succeed."

Just as her husband had a way of working with the drivers, she had a unique way of working with 'Big Bill.'

"Let me share with you a one-time secret," her granddaughter continued on. "She kept two sets of books. She kept the set of books that were the real set of books for the business, and then she kept the set of books that she shared with my grandfather, Bill France, Sr., just to make sure that he didn't spend us out of business. I think everybody in this room today should be thankful for that."

France Kennedy, the daughter of Bill France Jr., is carrying on her grandmother's legacy and serves on NASCAR's Board of Directors as Vice Chair of NASCAR. Her involvement and role in shaping the sport's future has been paramount to its success. She oversaw the innovative and massive redevelopment of Daytona International Speedway's grandstand and fan experience in 2016, as well as the revitalization of Phoenix Raceway in 2018, among many things.

Growing up among some of the most influential and powerful women in the sport's early days, France Kennedy believes working from the ground up as a young woman allowed her the proper perspective that allowed her to succeed later in life.

"My first job was when I was twelve. I started working in the ticket office during vacations, holidays, weekends. I worked with my grandmother and a number of ladies there in the ticket office," she told Kyle Petty in 2021. "It was probably one of the best introductions because you got to work directly with the fans. I think it was probably one of the best marketing studies that you could do."

In today's NASCAR, women are prominent members of the media, public relations, logistics, and management of the sport. Women have made an impact in the garage area, as well, serving as mechanics, engineers, team owners, and as NASCAR officials.

In February 2015, Kim Lopez became the first female and first Hispanic chief starter of the Daytona 500 when she waved the green flag for the 57th annual Great American Race.

"I like being the role model, especially for young kids," said Lopez. "It shows them that if you put your mind to something, you can do whatever you want, whether you're a boy or a girl.

"In this sport, everything has been male dominated, but nowadays you see more females changing tires, working on the cars, engineers, and flagmen, so it's great."

With the NASCAR Next and the Drive for Diversity programs, the next generation of female racers has a clearer path to follow into the sport. Just like any driver, though, it is up to them as racers to make their mark, attract sponsors, and succeed to the next level.

While today's NASCAR has more female involvement, they were hardly the first. Women racers are as old as the sport itself.

Sara Christian, the wife of an Atlanta, Georgia bootlegger, made her first start during NASCAR's first Strictly Stock race at Charlotte Speedway on June 19, 1949. Driving a Ford owned by her husband, Frank, Christian finished 14th out of 33 cars entered that day.

The first female NASCAR driver, Christian made a total of seven starts between 1949 and 1950. After finishing sixth in the season's fourth race, a 200-mile event around the one-mile Langhorne Speedway, NASCAR officials brought Christian to celebrate along with winner Curtis Turner. During the 1949 season, she recorded an average finish of 13.0, with one top-five and two top-tens. Entering six of the eight races that year, she finished outside the top-twenty just once and had a best finish of fifth at Heidelberg Raceway in Pittsburgh, Pennsylvania.

However, Christian was not alone in those early days.

The Daytona Beach and Road Course followed Charlotte Speedway as the second race of NASCAR's first Strictly Stock season. On July 10, 1949, three women entered the 28-car field for NASCAR's second Strictly Stock race.

Ethel Flock Mobley, the sister of Fonty, Bob, and Tim Flock, finished the highest of the three in 11th, followed by Christian in 18th, and Louise Smith in 20th. This also marked the first time a brother and sister competed in the same NASCAR event. Mobley competed once more that season at Langhorne Speedway, her only two NASCAR Cup Series starts, although she had over 100 starts in the NASCAR modified series.

Smith had also made a name for herself in the modified divisions, gaining notoriety in the pre-NASCAR days of 1947 for flipping her car during the Daytona Beach and Road Course event, pictures of which were circulated throughout various news publications.

Smith attempted the third race of the year at Occoneechee Speedway in Hillsboro, North Carolina. However, during a practice run she flipped her No. 94 Smith Auto Parts Ford. A popular driver on the circuit, she posed for pictures with her wrecked car and was listed as 27th out of 28 cars in the race.

The Greenville, South Carolina, native would race in eleven NASCAR events between 1949 and 1952.

Women did not only have to put up with the competition on the track, but off the track there were a host of challenges as well. Females were not allowed in the pits during the sport's earliest days. According to NASCAR Hall of Fame historian Buz McKim, Christian, Mobley, and Smith often had to run from the grandstands into their cars at the beginning of races.

It was not until the 1970s until women were allowed in the pits. A female photographer threatened to sue NASCAR in 1973 if they did not allow women in the pits. Many tracks on the NASCAR circuit displayed "No Women Allowed" signs in the garage and pit area.

According to McKim, Darrell Waltrip's wife, Stevie, was listed as his car owner when he started racing. She also lobbied to allow women in those areas and became the first female to sit on his pit box during a race.

While there are currently no female drivers competing full time in the NASCAR Cup Series as of the 2023 season, women play a prominent role on pit road as engineers, tire specialists, pit crew members, and more.

Women were also involved on the ownership side of the sport from the very beginning. After catching the racing bug watching events at Valdosta "75" Speedway in Georgia, Betty Lilly did all she could to find a way to become involved with stock car racing. After a race at the local short track one evening, Lilly approached her favorite driver, Sam McQuagg, about driving in NASCAR.

"I looked over and there was a lady sitting in a wheelchair. I walked over and she introduced herself as Mrs. Betty Lilly," McQuagg told Clyde Bolton of the Birmingham (Alabama) News. "She told us how much she enjoyed the sportsman races and how sorry she was they were rained out. Then she asked me how much I would like to drive Grand National. I told her driving Grand National was the dream of every race driver. She told me to call her and we'd talk about it."

Lilly provided the money for a 1965 Ford, and together she and McQuagg went racing at NASCAR's highest division.

Lilly's first NASCAR race as a team owner came in the Daytona 500 qualifying races, where McQuagg finished fifth. In the 1965 Daytona 500, McQuagg drove the Betty Lilly–owned Ford to an eighth-place finish.

The driven female car owner would go on to compete in thirty-nine NASCAR events over the next three seasons with McQuagg, Bobby Allison, Darel Dieringer, Ned Jarrett, Tiny Lund, Curtis Turner, Jack Harden, and Bobby Mausgrover behind the wheel. Her best finish as a car owner came during the 1966 season with Allison behind the wheel. The future NASCAR Hall of Fame member and leader of the famed "Alabama Gang" finished third on two occasions that season at Rockingham Speedway and Middle Georgia Raceway in Macon.

Female car owners and team principals continue to play a major role in NASCAR and that role is ever-expanding.

Teresa Earnhardt, wife of Dale Earnhardt, helped found their NASCAR Cup Series program, Dale Earnhardt Inc., and continued the organization after his death in 2001. The organization earned twenty-four NASCAR Cup Series wins, including three Daytona 500s with Michael Waltrip (2001, 2003) and Dale Earnhardt Jr. (2004). In 2008, DEI merged with Chip Ganassi Racing to form Earnhardt Ganassi Racing, with Jamie McMurray winning the 2010 Daytona 500, Brickyard 400, and Bank of America 400 at Charlotte Motor Speedway under that ownership banner.

In 2022, Jessica McLeod was named CEO of Live Fast Motorsports, co-owned by B. J. McLeod, Matt Tifft, and Joe Faulk, one of the first women believed to hold that role for a NASCAR Cup Series team.

"Jessica has been a crucial asset to LFM since the start," Tifft said. "Her direction, hard work, and dedication to the team have served as foundational assets leading us to successes in our team's first year, 2021. Today as LFM starts our second-year racing in the Cup Series, it feels right to give credit where credit is due, by naming Jessica with the well-deserved title and promotion of CEO."

When Ricky Stenhouse Jr. delivered a Daytona 500 win for JTG Daugherty Racing in 2023, history was made once again as it marked the first time a female co-owner and an African American co-owner went to Victory Lane in the sport's biggest race. Husband and wife owners Tad and Jodi Geschickter made the jump to the NASCAR Cup Series in 2009 and have been an integral part of the sport as one of the smaller, independent teams ever since.

Within NASCAR itself in the 2000s, women have risen to more prominent roles than ever before. As of 2023, three NASCAR racetracks have female presidents, including Richmond Raceway, Phoenix Raceway, and the Chicago Street Course.

On June 23, 2005, former Daytona International Speedway Vice President Gillian Zucker was named as President of California Speedway, officially taking over the role on July 5 of that year. Zucker served in that role until 2014.

On November 1, 2022, Latasha Causey was appointed track president at Phoenix Raceway, taking over to role from Julie Giese, who served in that role since October 2018. While Giese would lead NASCAR's efforts at organizing its first ever street race, Causey was tasked with leading the track that annually hosts NASCAR Championship Weekend. By becoming track president, Causey became the first female African American to hold the position of track president.

While Danica Patrick may be the most recognized and accomplished female racer in NASCAR's premier division today, the women that came before her blazed her path to the top of the sport. However, she—along with a host of other women—are setting a new example and serving as role models to younger generations.

Women have always been an integral part of NASCAR, whether behind the scenes or behind the wheel, and with the talented crop of young drivers coming up the ranks, the possibility of a first-time female winner in NASCAR may be just around the corner.

The First Rookie of the Year
Award Winner

THE SUNOCO Rookie of the Year Award is handed out at the end of each season to the first-year driver that has gained the most points over the other first-year competition. The rookies compete not only in the overall driver standings, but also a separate set of points that are tallied at the end of the season to crown a Rookie of the Year.

There have been some fierce battles and elite rookie classes over the years, but they can all trace their roots back to William H. "Blackie" Pitt.

Pitt, a Rocky Mount, North Carolina native, earned the first Rookie of the Year honor in 1954. Prior to that season, NASCAR did not recognize the best rookie in the field. Starting with Pitt, the sanctioning body selected the top rookie, as no true rookie point system existed.

During the 1954 season, Pitt competed in twenty-seven of the year's thirty-seven races driving for team owner Gary Drake. Pitt drove the first nine races of the season in a Plymouth and the final eighteen in an Oldsmobile. He recorded six top-ten finishes and earned $1,715.

From 1954 until 1973, the Rookie of the Year honor was decided by the sanctioning body with no system in place. That changed after Lennie Pond was given the award over Darrell Waltrip.

The decision was controversial and led to the formation of a well-organized point system for determining the Rookie of the Year Award winner moving forward. While competing for the overall driver championship, the rookie contenders would also battle one another with a completely separate set of point standings throughout the season to determine the Rookie of the Year Award on merit and accomplishment, not by committee vote.

For some, winning the Rookie of the Year Award was the first step in a long and successful career. Drivers such as Richard Petty (1959), David Pearson (1960), Dale Earnhardt (1979), Rusty Wallace (1984), Alan Kulwicki (1986), Jeff Gordon (1993), Tony Stewart (1999), Matt Kenseth (2000), Kevin Harvick (2001), Kyle Busch (2005), Joey Logano (2009), Kyle Larson (2014), and Chase Elliott (2016) would go on to win premier series championships in their careers.

In 2017, Erik Jones became the first driver in NASCAR history to win Rookie of the Year awards in all three national touring series. In 2022, Austin Cindric became the first driver in NASCAR history to win the Daytona 500 and the Rookie of the Year Award in the same season.

Each season, the rookie class takes on a completely different look as the fresh faces attempting to make it in NASCAR's top division continue to change. Some years, the competition for the Rookie of the Year Award is fierce and tightly contested. However, in some cases, a lackluster field of rookie candidates led to some surprise award winners, such as Kevin Conway in 2010.

Stewart holds the record for most wins of any Rookie of the Year award winner, with three victories in 1999. Jimmie Johnson also recorded three wins as a rookie but lost out on the rookie title to Ryan Newman in 2002.

Blistering Fast: The First 200 Miles Per Hour Qualifying Lap

WHEN IT comes to NASCAR, it's all about speed and power. Perhaps no time was that better on display at the superspeedway tracks of Daytona International Speedway and Talladega Superspeedway than during the 1980s.

Emerging from a decade in which the sport grew to national prominence and the horsepower continued to increase, speeds on NASCAR's largest track continued to climb.

Running unrestricted carburetor engines, the pole speeds at Daytona and Talladega were well over 190 mph as the 1970s came to a close, and the 1980s ushered in a new era in the sport. It would quickly become an era of fast cars, high speeds, and qualifying records.

On April 29, 1982, Benny Parsons would break a threshold long approached, but never before met in an official qualifying event. During time trials for the Winston 500 at Talladega Superspeedway (then known as Alabama International Motor Speedway), Parsons's Waddell Wilson–prepared No. 28 Pontiac became the first car in NASCAR history to break the 200-mile per hour barrier during a qualifying lap at an average speed of 200.176 mph.

The lap was the fastest qualifying speed to date and broke the previous track qualifying record of 199.658 mph, set by former series champion Bobby Isaac in April 1970.

All told, twenty-nine of the forty cars entered in the 1982 Winston 500 recorded a qualifying lap in excess of 190 mph.

Parsons's qualifying lap was of little surprise that weekend, however. During his practice run prior to qualifying, Parsons's No. 28 Pontiac Lemans posted a lap of 200.419 mph, and his first qualifying lap on the 2.66-mile superspeedway was actually the first on record to break the 200-mph barrier at 200.013 mph. His second lap was faster and set the mark for years to come.

"When I saw the crowd standing up in the stands and clapping, I knew I was the fastest. The next question was did I go 200?" Parsons said. "I guess I'm a little surprised to get it. This is a brand new car, with a chassis design that I've never driven before. I thought I'd probably do about what Darrell (Waltrip) ran. But I was faster by one-tenth of a second and that's the difference between 199 and 200."

The record-setting lap also caught the attention of Parsons's competition as well. Longtime team owner and NASCAR Hall of Fame member Leonard Wood was certainly impressed with the 200-mph lap.

"It sure sounds a whole lot more impressive than 199.999," Wood, whose driver Neil Bonnett ran the 14th-fastest lap of the qualifying session, said. "And 199-anything is awfully impressive all by itself. So, 200 miles per hour? Ummm!"

While the competition was left scratching their heads in awe, Parsons knew his 200-mph lap was something special for the sport of stock car racing.

"I'm sure there'll be a lot of conversation about this in the next few weeks," he said. "It'll probably get bigger than it is. I feel good about this, because I'll probably be remembered for it."

Parsons's record held throughout the 1982 season, despite a close call from Geoffrey Bodine on July 30, 1982, at Talladega Superspeedway when he posted an average qualifying lap speed at 199.400 mph. When the NASCAR premier series returned to Talladega, Alabama, Cale Yarborough would not only set a new qualifying record, he would smash it.

Driving a Harry Ranier-owned No. 28 Chevrolet, Yarborough posted a lap at 202.650 mph.

Parsons's qualifying record was not the first official lap over 200 miles per hour, however. That honor goes to Buddy Baker, who accomplished the feat at Talladega Superspeedway during a test session on March 24, 1970, twelve years before Parsons's record lap.

The test came about less than a year after the massive 2.66-mile facility opened and was deemed a Chrysler transmission durability test. Baker, who often served as a test driver throughout the1970s, drove a Dodge Charger Daytona throughout the day as engineers tweaked the car.

During that afternoon, Baker posted an average speed of 200.096 mph on the 30th lap of the session. This marked the first time a NASCAR stock car had broken the 200-mph barrier in the history of the sport.

Later in the session, the four-time Talladega Superspeedway winner ran his fastest lap of the session at 200.447 mph. Baker described it as "the most wonderful feeling," and said the accomplishment of being the first driver to run 200 mph on a closed course circuit was something no one could take away.

"It's hard to explain exactly what your train of thought does in a situation like this," said Baker. "You have to completely divorce yourself of any other thoughts—you become part of the race car. It surprised me that a car running that fast could be as stable as ours was. I was in the bottom lane all the way around."

Parsons and Baker had opened to door to the 200-mph barrier, and throughout the 1980s drivers such as Yarborough and, most notably, Bill Elliott would continue to increase the overall speeds and set blistering records in their wake.

Arguably one of the best superspeedway racers in NASCAR history, Elliott currently holds ten of the top twenty-five fastest qualifying laps in NASCAR premier series history, including eight of the top ten.

On April 30, 1987, Elliott ran the fastest recorded lap in a NASCAR premier series qualifying run at Talladega Superspeedway with an average speed of 212.809 mph. During that same qualifying session, each of the top twenty cars ran in excess of 207 miles per hour.

In that weekend's Winston 500 on May 3, Bobby Allison's car blew a right rear tire, was sent flying into the air, and ripped out a 100-foot section of the protective catchfence just ahead of the flag stand. While Allison was not hurt in the incident, one spectator was hospitalized and others treated for minor injuries from flying debris.

Following that frightening incident, NASCAR implemented the use of restrictor plates on the superspeedway tracks of Daytona and Talladega, restricting the airflow into the carburetors and thus reducing overall horsepower and speed.

As a result, qualifying speeds would never again reach the threshold set by Bill Elliott in the mid-1980s.

As NASCAR introduced the Generation Six car in 2014, qualifying speeds would increase on non-restricted tracks, primarily Michigan International Speedway.

On August 15, 2014, four-time premier series champion Jeff Gordon posted an average qualifying lap of 206.558 mph at the two-mile D-shaped oval, the seventh fastest qualifying speed in NASCAR history. On that day, thirty-two of the forty-three cars that made a qualifying attempt did so in excess of 200 mph.

"It doesn't feel that fast, to be honest." Gordon said. "It feels fast in the corners because I know I've got a tremendous amount of throttle in it and I'm carrying a lot of speed, at the same time when you're carrying that much speed in the corners you don't feel like you're going that fast on the straightaway. It's not a big acceleration or change. The only thing that gets my attention is when I look up there [at the scoring tower] and I see the speeds. You don't realize you're going that fast."

So, while NASCAR drivers continue to push past the 200 mph mark, they can all look to 1973 premier series champion Benny Parsons as the first to break the threshold and taste success at 200 mph in qualifying.

The First Live Flag-to-Flag Television Coverage of the Daytona 500

EACH YEAR, the NASCAR season kicks off in a big way at Daytona International Speedway for the "Great American Race," the Daytona 500. Unlike most sports, NASCAR starts the season off with the biggest race of the year, one that every competitor wants to win and every fan wants to experience. The event has grown into a major spectacle that is broadcast into the homes of millions of people around the world.

However, that was not always the case. In order for it to become the made-for-TV event it is today, NASCAR had to convince CBS Sports in 1978 that it was worth the airtime.

NASCAR's growth in the 1950s ushered in more attention from the media, and starting in 1954, television became interested in Bill France Sr.'s stock car racing circuit. During the 1954 Daytona Speedweeks, WABD-TV out of New York City, New York, carried a half-hour racing program called *Wire Wheels*, which focused the entire program on NASCAR's events in Daytona.

As the sport continued to grow, CBS Sports took notice and sent a fifty-man crew to Daytona International Speedway in 1960 to broadcast live Grand National qualifying races for the Daytona 500, as well as two Compact Car races. The races aired on January 31, 1960 on the two-hour *CBS Sports Spectacular.* It marked the first time a NASCAR race had been broadcast live on television.

NASCAR would take another step toward the future on April 10, 1971, when ABC's *Wide World of Sports* broadcast the first live flag-to-flag television coverage of a race. NASCAR and ABC Sports had agreed to televise the 100-mile race at Greenville-Pickens Speedway in South Carolina live on the guarantee the race would fit into a ninety-minute window. Bobby Isaac dominated the event that was slowed by just one caution flag. Thus the race took just one hour and sixteen minutes and ended fourteen minutes before the broadcast was scheduled to come to a close.

Looking to grow the sport and take it to the next level, Bill France Jr. enlisted the help of broadcaster Ken Squier and fast-talking driver Darrell Waltrip to convince Neal Pilson, President of CBS Sports, that the Daytona 500 was worth the network's airtime.

"They had to come see it. They really had to get a feeling of what it was and the flavor of it," said Squier, who served as announcer for the Daytona 500 from 1979 to 1997. "It was very hard, because a lot of those folks were city people, and they didn't grasp what it meant to the breadbasket of America, middle America, or to New England or to the South or the Northwest. Wherever you went in this country after World War II, there was that explosion of energy.

"That part of it, that middle ground of America—and those were the folks that voted with the ratings when that race was over—that was their game. People that lived in the cities, it was a little more difficult to understand cars running around about three or four blocks [in distance] and running into each other," he said. "They discovered there was a whole lot more to not only the race, but the people involved. They were regular folks, they didn't think they were anything special. I always thought that was one of the biggest selling points of stock car racing, common folks doing uncommon deeds."

Bill France Sr. had a way of doing uncommon deeds, and with the help of Squier and Waltrip, CBS Sports agreed to the deal—with a few conditions. In order to ensure full grandstands—something essential to NASCAR since the earliest days—France insisted the race be blacked out in the surrounding states.

When tickets flowed from the sales offices on race morning, France agreed to lift the blackout on the surrounding states (with the exception of Florida) so that the Southeast was able to get in on the historic action that was to follow.

On May 16, 1978, the partnership was announced to the public. With the deal finally done, the group celebrated with a dinner at Steak 'n Shake in Daytona Beach, Florida.

To shoot the massive event, CBS Sports brought in the same television crew that was used to broadcast Super Bowl XIII a month earlier. Michael Pearl was the director and Bob Fishman served as the producer.

The event saw the first use of a television camera along the outside retaining wall, as well as the first use of an in-car camera. The videographer along the wall stood boldly behind his camera at the exit of Turn 4, wearing a helmet just in case.

With unique pictures coming from all around the track, Benny Parsons provided viewers with a view never seen before. Parsons became the first driver to carry an in-car camera in a NASCAR race. In today's NASCAR, each driver carries multiple cameras inside of the car.

While it took some convincing of the CBS executives, there was no doubt in the minds of those involved that the broadcast would be a landmark event.

"It was just a matter of time before someone did it," Squier said. "We were fortunate that CBS gave us the opportunity to do it, and to do it right. They were willing to put up the dollars to really make something special of it—it wasn't just another sporting event—and to treat it with its proper respect. That was very, very special."

They say even the best-laid plans run into unforeseen issues, and this was no exception. After months of planning and meetings, Mother Nature wreaked havoc on the morning of the race. When crews, drivers, and fans arrived Sunday morning, rain was soaking the two-and-a-half-mile superspeedway.

Unable to race in the rain with slick tires, NASCAR and Bill France faced a potential disaster with rain pounding the track. Never to be deterred, France stood his ground and guaranteed the race would go on as scheduled.

"It rained in the morning, and we didn't know if we were going to get on the air," Squier recalled. "[Longtime motorsports reporter] Dick Berggren was walking across the infield and he heard Bill France Sr.'s voice saying, 'Well, the skies are going to clear and we're going to have a great race, we're going to be a little late getting started and thank you all for coming.' Dick said his tennis shoes were sinking into the water in the infield. He said, 'Honest to God I looked up and sun was coming through.' France Sr. could get it done."

Not only did France get it done, the potentially disastrous situation turned into a major advantage. While Daytona Beach was hit by rain, the majority of the East Coast was snowed in after a blizzard struck over the weekend.

With a large section of the country snowed in, they had few entertainment options. Luckily for France, CBS and the future of the sport, full flag-to-flag coverage of the Daytona 500 was one of those few options.

The snow opened the door to a host of non-traditional NASCAR fans, many of whom may have never tuned in otherwise. By the end of the day, many would never miss another race broadcast and the course of NASCAR history would never be the same.

Concerned about maintaining the contract it signed with CBS and to ensure television money in the overall race purse, NASCAR did all it could to get the race started on time at 1:00 p.m. ET.

After being introduced to the crowd, drivers were told to climb in their cars and prepare to race despite the wet track conditions. At 1:00 p.m. ET, the field took the green flag under yellow conditions as the heat of the cars worked on drying the racing surface.

With the rest of the field still going slow under caution, Waltrip took his No. 88 Gatorade-sponsored car out and ran a few laps at full speed to test the conditions of the racing surface. When Waltrip said the conditions were fine to go racing, NASCAR made the decision to put the race under green flag conditions.

On Lap 16, the field took the green flag and sped off into the first turn to officially start the race. However, it would not take long for the leaders to create a bit of drama and bring out the caution flag yet again.

Fifteen laps after officially going racing, Bobby Allison, Donnie Allison, and Cale Yarborough would tangle off the exit of the second corner battling for the top spot.

Bobby Allison claims he was hit from behind, but either way, the NASCAR Hall of Fame member would wiggle off Turn 2 and make contact with his brother Donnie's left rear, spinning him down the track. Yarborough was running third and had nowhere to go but get collected. The trio of cars—running first, second, and third at the time—slipped and slid through the rain-soaked infield grass. The innocent incident helped to build the drama that would eventually explode after the race.

Luckily for CBS Sports and NASCAR, the cameras were trained directly on the three cars as they tangled off the second corner and slid through the grass in dramatic fashion. If the viewers were not hooked by the opening lap, this incident certainly had their attention.

"You look back and say, 'Hey, that's installment one of something that's going to be really good,'" said Jack Arute, Motor Racing Network's executive producer at the time.

As the race went on, it developed into a highly competitive event featuring thirty-six lead changes among thirteen different drivers. In the closing laps, however, it would come down to a fierce battle between Donnie Allison and Cale Yarborough.

Coming to the white flag, Yarborough's No. 11 Oldsmobile was tucked closely behind the No. 1 Oldsmobile of Allison. As the pair exited the second corner for the final time, few could have predicted what would happen next.

In full view of the television cameras and the viewers watching at home, Yarborough made a move to the bottom of Allison for the race lead. Allison threw a block and drove Yarborough down to the infield grass. The rear of Yarborough's car slipped, sent him up into Allison's, and started a series of events that would see the two cars bounce off each other before crashing back into one another.

As the two cars battled down the backstretch, they finally hooked bumpers and slid up the track into the wall as they entered the third turn. The contact damaged both cars and sent them sliding down to the infield grass, still hooked together and spewing smoke.

When the two cars came to a rest, the CBS broadcast team and the fans at home were left wondering who was in the race lead. "Who's going to win it?" Squier said on the broadcast as the cars finally came to a rest on the apron.

With no clear camera shot of Richard Petty, the third-place car that would inevitably inherit the lead, the producers and announcers were presented with an unanticipated and unpredictable event. The camera eventually picked up 12th-place Buddy Arrington, whose car was painted similar to Petty's. Squier immediately recognized the mistake, but made up for it with his amazing call.

"That's when we started selling the race from the booth," Squier said.

After a few moments of scrambled confusion, the cameras picked up the tight battle for third between Petty, Darrell Waltrip, and A. J. Foyt. As they sped past the wreckage of the leaders, it was quickly clear this was the battle for the win. Unable to capitalize on the moment, Waltrip—who worked hard to convince CBS officials to carry the race—would cross the start-finish line second behind Petty.

"From our perspective, this was amazingly good press," said Alexis Leras, NASCAR news bureau director. "The King of NASCAR, Richard Petty, wins. The first 500-mile race telecast in its entirety."

While the cameras remained focused on Petty and his victory parade to the winner's circle, Squier announced some of the most famous words in NASCAR history, "And there is a fight! Between Cale Yarborough and Donnie Allison. The tempers overflowing!"

As the cameras beamed images into the homes across the country, the Allison brothers were coming to blows with Yarborough on the infield grass.

After the cars of Donnie and Cale came to a rest, Bobby stopped to check on his brother. Cale went over to Bobby's car and blamed him for the wreck early in the race. Bobby climbed out of the car and it was on from there. Punches were thrown, and Bobby grabbed Cale by the ankle with one hand and the collar with the other, driving him to the ground.

Bobby enjoys telling the story by saying, "That's when Cale went to beating on my fist with his face."

The two Allison brothers fought Yarborough on the infield grass in the third turn as the television audience watched from a variety of camera angles, enthralled with what they saw.

After the race, when reporters asked Petty what he thought of the wrecking and fighting on the last lap, he thought back to his infamous run-in with rival David Pearson coming to the checkered flag in the 1976 Daytona 500. The two lead cars wrecked off the fourth turn and Pearson limped across the finish line to take the victory as Petty's car came to a rest in the infield grass.

"I said, 'Them cats don't have any class,'" Petty remembered. "They said, 'What do you mean?' I said, 'When me and Pearson did it, we did it over by the grandstand where everybody could see us. Cale and Donnie did it on the backstretch out of sight.'"

While the wreck and subsequent fight may have been out of sight from the fans in the frontstretch grandstands, the people watching at home on CBS saw every dramatic and entertaining moment unfold live.

Yarborough and both Allison brothers were fined for the incident, but their epic battle in front of a live national audience forever changed the course of history. NASCAR suddenly went from a regional sport with little attention to gracing the front page of the *New York Times* sports section the following day.

Today, NASCAR enjoys a lucrative television contract with FOX and NBC, who together broadcast nearly every practice and qualifying session for each of the sport's top three national touring series, and every single race is covered live flag-to-flag.

All of these years later, Squier remains adamant that France knew the gamble and risk of that first live nationally broadcast Daytona 500 would grow into the major powerhouse it is today.

"There wasn't any question," Squier said. "France Sr. told me when I first went to work for him, 'You understand that coming down here this sport will be one of the major league sports come 2000.' Well, he was wrong. He missed it by about six months."

Herb Thomas (left) was the first driver to earn three victories in the Southern 500 at Darlington Raceway, as well as the first two-time NASCAR premier series champion. (*AP Photo*)

Second-generation racer Richard Petty celebrates winning his 200th career NASCAR race during the 1984 Firecracker 400 at Daytona International Speedway. Petty became the first driver to win 200 races, seven championships, and seven Daytona 500s. (*AP Photo/Elliot Schecter*)

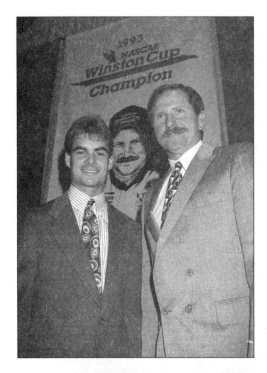

After Jeff Gordon's first start in the 1992 Hooters 500 at Atlanta Motor Speedway, NASCAR's new "Wonder Boy" went on to win the 1993 Rookie of the Year Award and impress Winston Cup champion Dale Earnhardt in the process. (*AP Photo / Ed Bailey*)

Following the death of Dale Earnhardt in the 2001 Daytona 500, NASCAR introduced a host of safety innovations, including the installation of Steel and Foam Energy Reduction (SAFER) barriers at almost every track. (*AP Photo / Paul Kizzle*)

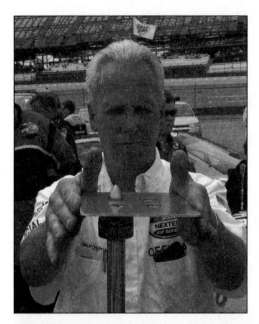

NASCAR implemented the use of restrictor plates for each car at Daytona International Speedway and Talladega Superspeedway in 1988, reducing the horsepower of the engine to slow down speeds at the high-banked tracks. (*AP Photo/ Gene Blythe*)

Lee Petty (center) was the patriarch of a family that would include his two sons, Richard (right) and Maurice (left), his grandson, Kyle, and great-grandson, Adam. Lee, Richard, and Maurice are all members of the NASCAR Hall of Fame. (*AP Photo*)

Ronald Reagan helped take the sport to a new level when he became the first sitting president to attend a NASCAR race, the 1984 Firecracker 400 at Daytona International Speedway. President Reagan helped call some of the race with Motor Racing Network's Ned Jarrett and celebrated with Richard Petty after he scored his 200th NASCAR win. (*AP Photo/Ira Schwarz*)

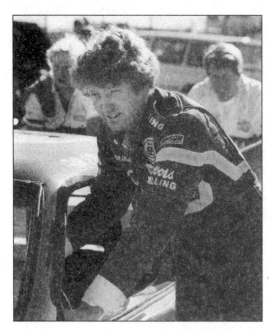

On April 30, 1987, Bill Elliott ran the fastest recorded lap in a NASCAR premier series qualifying run at Talladega Superspeedway with an average speed of 212.809 miles per hour. (*AP Photo/Lennox McLendon*)

Mike Skinner was one of the earliest stars of the NASCAR Camping World Truck Series and won the inaugural series championship in 1995. (*AP Photo/Denny Medley*)

Glory Road at the NASCAR Hall of Fame in Charlotte, North Carolina showcases some of the most historic cars in NASCAR history, as well as the varying degree of banking at each NASCAR track. (*AP Photo/Chuck Burton*)

Seven-time NASCAR champion Richard Petty was among the first class of the NASCAR Hall of Fame. Bill France, Bill France Jr., Junior Johnson, and Dale Earnhardt were also part of the inaugural 2010 class. (*AP Photo / Chuck Burton*)

Jimmie Johnson won an unprecedented five consecutive NASCAR Sprint Cup championships from 2006 to 2010. The Hendrick Motorsports driver also won the 2013 championship, moving him within one of the all-time mark of seven titles held by Richard Petty and Dale Earnhardt. (*AP Photo / Lynne Sladky*)

Roof flaps were implemented by NASCAR for the start of the 1994 season in an effort to reduce airborne crashes, like the two that involved Rusty Wallace at Daytona and Talladega in 1993. (*AP Photo/Garry Jones*)

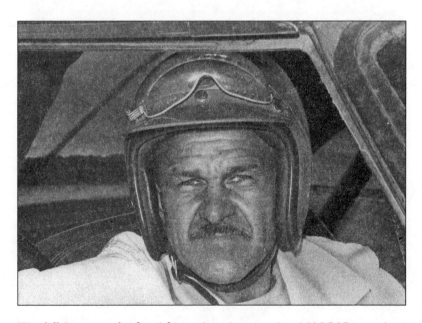

Wendell Scott was the first African-American to win a NASCAR premier series race, but it was not without controversy. In 2015, Scott was inducted into the NASCAR Hall of Fame. (*AP Photo/File*)

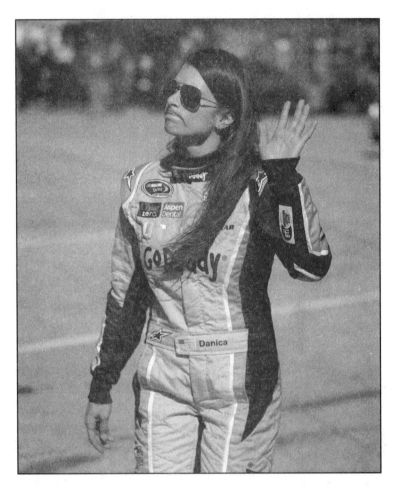

Danica Patrick follows the legacy of the female racers throughout NASCAR history, and in 2013 she became the first female driver to qualify on the pole for the Daytona 500. (*AP Photo/John Raoux*)

The King Meets a President

As NASCAR grew through the modern era in the 1980s, the sport became more nationally recognized. The first flag-to-flag nationally broadcast Daytona 500 in 1979 was full of action, drama, and entertainment, and as the 1980s continued toward the midpoint of the decade, NASCAR was becoming more and more a part of the American conversation.

Perhaps there is no better an example of that impact as when Ronald Reagan became the first sitting president of the United States of America to attend a NASCAR event.

And boy, did he pick a good one. The 40th US President flew on Air Force One to Daytona Beach, Florida for the July 4, 1984, Firecracker 400 at Daytona International Speedway.

The discussions about Reagan's trip to Daytona began in the weeks leading up to the race when NASCAR team owner and National Finance Committee Chairman Mike Curb was sitting in on a meeting in which the president's Fourth of July plans were being discussed.

President Reagan was scheduled to lay a commemorative wreath at Arlington National Cemetery early in the day but was open to suggestions for additional plans. Curb would provide a suggestion that would shock many in the room, but ultimately lead to one of the most memorable moments in American sports history.

"I said, 'Mr. President, you used to love calling auto races when you were in Iowa. What about going down to Daytona?' He said, 'I love NASCAR.' By then, it had been on TV for a few years. We talked about the '79 Allison-Yarborough fight," Curb recalled. "He said, 'Who are those two guys who got in the fight?' I could tell from talking to him that he knew about the race. He didn't mention them by name, but he remembered the incident.

"So he had been watching NASCAR to some degree. I said, 'Why don't you come to Daytona? The airport's right next to the speedway. They'd probably even let you call the race if you wanted to.'

"The people in the room were horrified. They thought I was crazy. Then all of a sudden, President Reagan said, 'You know, I think I would enjoy that.'"

In the days leading up to the event, Secret Service members scoured the area ensuring security would not be an issue. The agents put in charge with ensuring the president's safety searched under the track in manholes, posed as food vendors, and took up residence in a local hotel.

On the day of the race, Reagan completed his obligations in the Washington, DC, area before boarding Air Force One for the trip to Daytona Beach, Florida. While on the presidential plane, Reagan lifted a telephone and gave the command to fire engines from the front of Air Force One.

"We could hear him, they put him on the loudspeaker. Everyone was setting in the car waiting to go," said Petty. "It was a fuzzy day. There was so much happening, so much going on— July the Fourth, the president announced crank your engines. After that was over with, we forgot about the president. The racers were just out racing. We didn't realize what a big deal it was going to be."

Air Force One landed at Daytona International Airport behind the speedway as the race was going on. A photographer with impeccable timing captured the president's plane coming in for landing as Richard Petty was driving his famous No. 43 out of the second corner. The image would be just one of the iconic moments to emerge from this historic day.

Once on site, President Reagan made his way to the broadcast booth, joining Motor Racing Network (MRN) to help call some of the action. Sitting with NASCAR Hall of Fame member and MRN color commentator Ned Jarrett in the broadcast booth, President Reagan was captivated by the speed of the cars and the ease Jarrett had in calling the high-paced race. In turn, Jarrett urged the president to call a few laps, which he struggled through not knowing who was in what car.

"It was one of the highlights of my life to be chosen to do that and get to spend that time with the president," said Jarrett. "What a down to earth, neat person he was. He made you feel so at home. You'd think that you'd be nervous and forget what you were even there to do—forget about what was going on on the racetrack. But he just made you feel so at home. He fit right in. We just had a big time in there. I think it was while we were on the air, I know he mentioned to me one time, 'I understand that you have a son racing in this race.' That was Dale's first race on a big track in what was then the Winston Cup Series. He had been briefed pretty well on the situation."

After a lengthy stay in the broadcast booth with Jarrett, President Reagan made his way back up to the France suite to watch the final laps with Bill France Jr. on his left and Bill France Sr. to his right. No briefing could have prepared President Reagan for what was going to take place over the closing laps.

The race was shaping up as a classic battle between Richard Petty and Cale Yarborough. The two legendary drivers had the best cars all race long and were clearly the top cars in the field. While Yarborough was trying to sweep the season's races at the 'World Center of Racing' after winning the Daytona 500, Petty was searching for his monumental 200th career victory.

With the laps clicking away, Yarborough was glued to the back bumper of Petty's famous No. 43. An ace at Daytona International Speedway, Yarborough was planning a last-lap slingshot pass to get by Petty and score the win. Knowing the keen racer Yarborough was, Petty slowed the pace of the race by backing off the throttle a bit more each lap to save some of his equipment for the final run to the checkered flag.

"I made a mistake in that race," Yarborough recalled. "My car was quick enough that I could have won the race instead of waiting for that last-lap move."

Instead, Yarborough remained tucked behind Petty as he waited for the perfect moment to make the pass and score the win. Yarborough would never get that chance.

As Petty and Yarborough were planning for the final lap showdown, the No. 01 Chevrolet of Doug Heveron went for a wild slide, exiting the tri-oval with his car lifting into the air once it hit the infield grass. The incident brought out the caution, but the two leaders had already passed the start-finish line when incident occurred, meaning the first car to the yellow flag would win the race.

"I always say, if I didn't make it, at least I'll be a trivia question," Heveron later said. "There were three or four laps to go. I was pulling out to pass [another driver], and Benny Parsons was in my blind spot. I didn't know he was there. We didn't really have the best spotter. I pulled out, and he just caught my right rear bumper. It got me a little wormy. Of course, it went left, and I was skidding sideways. When the windshield left, I said, 'Well, this isn't going to be a whole lot of fun.' The thing took off and just went up and landed on the driver's door. I had mud inside my helmet. I had an open-faced helmet then. There was mud everywhere. Thank God it rained, because I believe I hit my head on the ground."

All the planning for that final lap immediately went out the window as Petty and Yarborough raced hard into the first corner. Understanding the drama and intensity that was unfolding on the track, President Reagan rose from his seat and watched intensely as the final moments of the race unfolded in front of his eyes.

While Yarborough was able to get a run on Petty down the backstretch and make a move for the lead into Turn 3, the car drifted up the banking and allowed Petty to dive underneath and pull side-by-side. Coming off the final corner, the two cars were even with one another as they closed in on slower traffic up ahead. The two cars made slight contact in the tri-oval as Petty took the caution flag about one foot ahead of Yarborough to his outside.

However, the race was not yet over. While the green-flag portion of the race had come to a conclusion, the field was required to stay on track and maintain pace car speed until the checkered flag flew. As Petty drove through the tri-oval to take the white flag under the caution period, Yarborough pulled his car to the pits. Realizing his mistake, Yarborough drove back onto the track but had lost the second spot to Harry Gant.

"I misread the flagman's fingers and when we came down to the caution, I thought the race was over," Yarborough said after the race. "I guess my brain blew up is what happened. It was just a dumb thing to do. I came in and lost positions."

Taking the checkered flag under yellow conditions, Petty had earned his 200th career NASCAR premier series victory in dramatic fashion on the track and on a day that would always live in history off the track.

After racing 400 miles on the track in the July Florida heat, Petty stopped his car at the start-finish line, climbed from the car with a wet rag in his mouth (Petty often raced with wet rags to keep cool throughout the race), grabbed his hat and sunglasses, and immediately ran up the banking, through the grandstands and up to France's suite to meet President Reagan.

For security reasons, Victory Lane celebrations were moved into the broadcast booth to allow Petty to celebrate his win with the president. Once the longtime Republican supporter made his way to the booth, Petty stood side-by-side with President Reagan for an interview with ABC Sports's Jim Lampley.

"When Reagan was in the room, he was the star. Not just because he was the president. He had a tremendous presence. But when he was there with Petty, you got the sense there were two stars in the room," recalled Chris Wallace, NBC White House Correspondent at the time.

Once the pomp and circumstances were complete, President Reagan joined the NASCAR community as they celebrated the Fourth of July holiday with a picnic of Kentucky Fried Chicken. President Reagan sat with Petty to his right and Bobby Allison to his left. The picnic tables complete with small American flag centerpieces and buckets of KFC.

The small southern sport organized and built by the France family had grown from the dirt tracks and dust bowls to superspeedways and live television broadcasts. Now, the entire sport was sitting down with the president of the United States of America to celebrate the Fourth of July holiday after putting on one of the most iconic and historic races in NASCAR history.

Once again, the eyes of the nation were turned to NASCAR and the sport delivered in an impeccable way. Much like the thrilling live finish of the 1979 Daytona 500, NASCAR proved it was home to some of the most talented drivers in the world and some of the best sports entertainment around.

"When you watched the news that night and saw the pomp and circumstance that the world attached to it, all of it, thirty years later, you're still sitting here saying, 'Wow, I can't believe all of that happened,'" said Mike Helton, former NASCAR President and Current Vice Chairman. "When he came to Daytona that day, [President Reagan] brought the world with him, for us. That's something you keep grinning about."

The First All-Star Race

EACH YEAR prior to the running of the Coca-Cola 600 at Charlotte Motor Speedway, the biggest names in NASCAR show off their skills in a non-points paying all-star race. Created in 1985 as a way to increase publicity for R. J. Reynolds Tobacco Co. (RJR), title sponsor of NASCAR's premier series from 1971 until 2003, the NASCAR All-Star Race has become one of the most exciting and highly anticipated events on the schedule.

RJR's version of the all-star race was not the first of its kind. From 1961 until 1963, Bill France Sr. hosted three all-star events at Daytona International Speedway. The American Challenge Cup, as it was called, featured the previous year's winners.

Joe Weatherly won the inaugural running of the event in 1961 from the pole after a heated battle with Herb Thomas and Junior Johnson, with the margin of victory listed as just 36 inches. Eleven cars competed in the 10-lap event around the two-and-half mile Daytona track. Fireball Roberts would go on to win the 1962 running of the all-star event, while Fred Lorenzen won the 1963 race, renamed the "Race of Champions." After three years, France's version of an all-star race came to an end.

Serving as NASCAR's title sponsor since 1971, RJR was looking for a better way to showcase their support and gain additional media coverage during the mid-1980s. RJR executives Ralph Seagraves and T. Wayne Robertson developed the

Winston Million, a one-million-dollar prize to any driver that could win three of the four marquee events on the Winston Cup schedule. However, they were still looking for more.

During the 1984 awards ceremony at New York City's famous Waldorf-Astoria Hotel, RJR president Gerald H. Long announced the title sponsor would host an all-star event during the 1985 season. The non-points exhibition event would take place at Charlotte Motor Speedway, feature the 1984 race winners, was scheduled for 105 miles, with the winner taking home $200,000. Thus, The Winston was born.

"R. J. Reynolds is committed to ensuring that Winston Cup racing has no rival when it comes to being first in motorsports," said Long. "The series is already first worldwide in attendance and prestige. This is another big leap forward. The Winston will be the richest per-mile race in the world."

The race was held as part of a doubleheader with a Late Model Sportsman division race, the day before the annual 600-mile event. The twelve race winners from the 1984 season were entered in the 105-mile exhibition event. The field included Terry Labonte, Darrell Waltrip, Harry Gant, Bill Elliott, Geoffrey Bodine, Cale Yarborough, Dale Earnhardt, Bobby Allison, Richard Petty, Ricky Rudd, Tim Richmond, and Benny Parsons.

Once the green flag dropped, Labonte, Waltrip, and Gant fought for the lead amongst themselves over the 70-lap event. In the closing laps, Waltrip was able to get around Gant for the lead and would hang on to take the historic victory. As Waltrip crossed the finish line to take the checkered flag, his motor expired and sent white smoke billowing behind his No. 11 Chevrolet.

Waltrip's team owner Junior Johnson had a direct connection with RJR and was determined to win their first all-star race and the $200,000 that went along with it. Johnson had been working with RJR on sponsorship in 1970 when he put them in touch with Bill France and NASCAR. While the team owner was simply looking for season-long sponsorship, RJR had a much larger budget and Johnson knew France could put it to use much better than a race team. When the planning began for The Winston in 1984, Johnson was once again heavily involved in the process.

"That race, Junior had a lot to do [with it] working with Ralph Seagraves and T. Wayne Robertson, and all the people at R. J. Reynolds," said Waltrip. "They have this vision for an all-star race like that. Junior was determined to win that race. We built a special car, just like they do today. We took it to the wind tunnel a number of times, which was unique. We didn't do that a lot back in the day. We tested that car at Charlotte. Junior Johnson built that engine.

"Junior said, 'This isn't going to last long. You're going to have to watch practice laps.' He said, 'This thing's going to run a couple hundred miles, that's it.' The rods were light, the pistons were light, everything in the engine was light," Waltrip added. "When it blew up, it shocked me, because I didn't expect it, but it didn't surprise me once it was over with."

While it may have shocked Waltrip, the perfect timing of the engine failure after crossing the finish line raised many questions and speculation about whether he had blown the motor on purpose to hide Junior Johnson's unique modifications.

When Johnson and Waltrip decided they wanted to race the car used to win The Winston in Sunday's World 600, NASCAR would not allow it. Waltrip had qualified fourth in another car, but after winning the all-star race, the team put the engine from the qualifying car into the all-star race car and prepared to race it during the 600-mile event. However, NASCAR Winston Cup Director Dick Beaty informed the team it was not permitted to change cars overnight.

"We had this car ready to go on the line, and they came and said, 'You can't race that car. You have to race the car you qualify,'" Waltrip remembered. "Junior said, 'Load it up, we're going home. There's no way we can change the engine back and put it in the car and get it on the line in time to make the race.'"

Cooler heads eventually prevailed and the team went to work swapping the engine back into the original car and getting it prepared in time for the start of the race. In the end it worked out well for Waltrip and Johnson, as they would go on to win the World 600 that day, with Waltrip becoming the first driver to win the All-Star Race and the 600-mile event in the same year.

"It turned out to be motivation," Waltrip said. "We were determined to win that race on Sunday just like we did on Saturday. It was a crazy weekend."

The following year in 1986, The Winston was moved to Atlanta Motor Speedway, with local favorite Bill Elliott taking the victory. The event returned to Charlotte Motor Speedway for the 1987 season and remained the home for the NASCAR All-Star Race until 2020.

During the 2020 season, the NASCAR All-Star Race was moved to Bristol Motor Speedway under the lights on a Wednesday in July. Due to COVID-19 restrictions, fans were not allowed to attend the event in person. Given the exhibition style event, NASCAR also allowed the team to add under glow lights to their cars, something never done. Chase Elliott scored the victory over Kyle Busch by less than half a second.

For the 2021 and 2022 season, the NASCAR All-Star Race shifted to another Speedway Motorsports Inc. track, this time Texas Motor Speedway. In a shuffle to get Circuit of the Americas on the NASCAR schedule, Texas forfeited one of its point-paying dates in return for the annual all-star event. Kyle

Larson captured the first NASCAR All-Star Race in Texas, while Ryan Blaney scored his only win of the season in 2022 in the non-point paying event.

A renewed interest and excitement for the NASCAR All-Star Race came about when it was announced the 2023 event would take place at the historic North Wilkesboro Speedway, a track returning to the schedule for the first time since 1996.

"The NASCAR All-Star Race has always been the fans' race, and I can't think of a better way to celebrate the fans during NASCAR's 75th anniversary than returning to North Wilkesboro Speedway," said Speedway Motorsports President and CEO Marcus Smith. "We couldn't have made this happen without tremendous support from so many people including Governor [Roy] Cooper, our state legislature, NASCAR, Dale Earnhardt Jr., and of course, the people of Wilkes County. We've got a lot of work to do, but we've got the will to create something special alongside a community and hardworking staff that will get it done."

Smith and his team went to work improving the long dormant facility, upgrading grandstands and facilities, but they left the racing surface alone. On March 21, 2023, NASCAR Cup Series cars officially returned to North Wilkesboro for the first time since September 29, 1996, as three teams—23XI Racing, Richard Childress Racing, and Roush Fenway Keselowski Racing—took part in a Goodyear tire test ahead of the NASCAR All-Star Race.

The amount of history in the facility and the racetrack was not lost on Richard Childress Racing driver Austin Dillon, who was part of the tire test.

"The track's still the old track. It just teaches you short-track racing [techniques], being easy on that gas pedal and trying to take care of the rear tires. We had a blast and I can't wait to come back," said Dillon. "Just finally getting here was cool. This place was dormant forever and now it's alive, and you can feel the energy and the momentum that's building. It's going to be electric [in May]. It's going to have that feel they had in the 1990s here, when they were packed out and fans were getting to enjoy a race near home. It's rough out there. It's gritty. It's North Wilkesboro."

The revival of North Wilkesboro also expanded the all-star weekend to include much more racing action than in previous years. With a renewed focus on short-track racing, the NASCAR All-Star Race week featured multiple late model stock car events from the ASA STARS National Tour and CARS Tour, which is co-owned by Dale Earnhardt Jr., Jeff Burton, Kevin Harvick, and Justin Marks. The weekend included the NASCAR Craftsman Truck Series race, Pit Crew Competition, and Heat races for the All-Star Race, which were run on wet weather tires.

Kyle Larson dominated the weekend at North Wilkesboro, winning both the Truck Series race and NASCAR All-Star Race. Leading 145 of the 200 laps of the All-Star Race, Larson was nearly unstoppable as he lapped all but twelve cars around the 0.625-mile short track. Taking the trophy and million dollars, Larson became the first driver to win the NASCAR All-Star Race at three different racetracks: Charlotte Motor Speedway (2019), Texas Motor Speedway (2021), and North Wilkesboro Speedway (2023).

"I didn't get the chance to race at Bristol, but to win at Charlotte was awesome. That's one of my first big wins in a stock car," Larson said after his North Wilkesboro win. "Then Texas was cool to win at a different track. Now you're like, man, they keep switching these things up. It would be cool to win at another venue at a historic venue like this.

"I am extremely proud for sure to have won the All-Star Race, but to win it at three different tracks is really, really neat," he said.

The NASCAR All-Star Race has created some of the most exciting and dramatic moments in NASCAR history and continues to be one of the most anticipated races of the season. Despite multiple name changes, locations, and formats, the action on the track continues to be some of the best around and the trophy is one all drivers want to win.

Speeding into the Night

THERE IS nothing quite like watching a NASCAR race at night. The colors shining bright and the light reflecting off the cars make the roaring spectacle a sight to see. While night racing is currently a common occurrence in modern NASCAR, the roots go deep into the sport's dusty past.

The first ever NASCAR race held under the lights took place on June 16, 1951, at the half-mile Columbia Speedway. Frank Mundy earned his first NASCAR victory by starting on the pole, leading twice, and beating Bill Blair by one full lap. The win was also the first victory for a Studebaker in NASCAR.

Later that year, another night race was held on Friday, September 7, also at Columbia Speedway. Tim Flock overtook his brother Bob Flock after 72 laps and never looked back. Driving a 1951 Ted Chester–owned Oldsmobile, the Hall of Famer led 128 of the 200 laps and earned $1,000 for the victory. Fireball Roberts was initially flagged the winner of the race, but when Flock's team called into questions the scoring, NASCAR reversed the decision and ruled Roberts had finished second.

Racing at night continued to spread, and the first night race run on a paved track came at the one-mile Raleigh Speedway in North Carolina on Saturday, May 29, 1954. That evening Herb Thomas won a total of $2,250 by leading 204 of 250 laps, two laps ahead of second-place finisher Dick Rathman.

NASCAR's premier series also raced under the lights at Fairgrounds Speedway in Nashville, Tennessee, starting with the Music City 200 on June 3, 1965. Dick Hutcherson won that night by leading every lap in a race in which only seven of the 15 cars entered finished the race.

As NASCAR grew into the modern era, night racing followed. The aesthetic allure of race cars at night, along with the tempers and sparks it often brought out, were perhaps best evoked when lights were added to Bristol Motor Speedway in 1978.

Cale Yarborough beat out Benny Parsons to win the Volunteer 500 on Saturday, August 26, 1978, becoming the first winner of the Bristol Night Race. The hard short-track action on the high banks, under the lights, created some of the best racing on the NASCAR circuit and quickly became a fan-favorite and must-see.

"They raced at Nashville at night, but when Bristol started doing it in the summer, that was by far the thing that brought the people out. Even back then, that night race was the one everybody wanted to come to," Yarborough would later say. "That night race was something special back in '78 but now, well, everybody knows how people love that [night] race in Bristol. And let me tell you, winning at Bristol is special . . . that's something I know a lot about . . . but winning that night race, that's something that you never forget. There's just something special about knowing you beat everybody with all those people watching at a place as tough as ol' Bristol."

In 1988, Richmond International Raceway was expanded from 0.542 miles to its current configuration of .75 of a mile. Three years later, lights were added ahead of the fall 1991 race. The Miller Genuine Draft 400 on September 7, 1991 marked the first time NASCAR raced at night at the newly configured track.

Harry Gant beat Davey Allison and Rusty Wallace to earn the victory, the second of a four-race win streak that earned him the nickname "Mr. September."

The next evolution of NASCAR night racing came in 1992 when Bruton Smith added lights to Charlotte Motor Speedway, making it the first major speedway to host a night race. The project cost Smith a total of $1.7 million and was the largest sports facility to add lights at the time. Never one to shy away from a publicity opportunity, Smith staged a media event to officially switch on the lights to the speedway, but all did not go according to plan. When Smith flipped the large switch, a shower of sparks came down upon him, briefly setting his hair on fire. Sparks and drama would come to epitomize the first night race at the one-and-a-half-mile speedway.

The 1992 Winston all-star race was dubbed "One Hot Night," and the race lived up to every expectation. The excitement in the air was palpable. Drivers were excited, fans were thrilled, and the race was being broadcast live on The Nashville Network. On a pace lap the field stopped in the tri-oval for a photo opportunity, as flashbulbs burst throughout the grandstands.

Once under green, The Winston was one for the ages.

The race's final segment came down to a battle between Dale Earnhardt and Kyle Petty. The Intimidator led on the white flag lap, while Petty got a huge run off the second corner and made a dive to the bottom for the lead. Earnhardt threw a block and drove his shining black-and-white No. 3 Chevrolet down to the dust of the backstretch to protect the lead.

As the cars raced through the third turn, Earnhardt's car lost the rear end and slid up the track, his black Chevrolet sending white smoke billowing. Petty raced below and opened the door for Davey Allison to make a move.

Allison's No. 28 Ford got the advantage through the tri-oval and took the win, but the two cars made contact past the start-finish line. Allison's car was sent hard into the outside wall, hitting the driver side in a shower of sparks, fireworks to end an historic evening.

Even the iconic Daytona International Speedway made the move to night racing in 1998. A massive 2.5-mile facility, Daytona became the largest lighted sports complex in the United States, with 1,932 light fixtures that could stretch from the speedway to Musco Lighting headquarters in Muscatine, Iowa.

Dale Earnhardt flipped the switch to turn on the lights for the first time in February 1998 in front of nearly 20,000 fans in attendance. The Daytona ace then ran 20 laps under the newly lit track.

For the 1998 season, the annual Fourth of July 400-mile race was moved to Saturday night under the lights—at least that was the plan. However, wildfires throughout the Central Florida area forced NASCAR to postpone the Pepsi 400 until October 17. Jeff Gordon won the rescheduled event, one of his thirteen victories that season.

Today, night racing remains a thrilling and aesthetically pleasing part of NASCAR, with the sparks flying and the tempers often flaring under the lights.

A Day Like No Other: Jeff Gordon's First Race

THERE ARE certain races in NASCAR legend that hold a special place in the history of the sport.

The inaugural Daytona 500's three-wide photo finish. The 1976 Daytona 500 finish with Richard Petty and David Pearson crashing as they came to the checkered flag. The 1979 Daytona 500 in which Donnie Allison and Cale Yarborough wreck on the final lap and fight after the race. Dale Earnhardt's emotional win in the 1998 Daytona 500 and his amazing charge to win the 2000 Winston 500 at Talladega.

However, perhaps none of them had quite as much significance, drama, and historic importance as the 1992 Hooters 500 at Atlanta Motor Speedway. The final race of the 1992 season was one of the most dramatic in the sport's history, as six drivers had a chance at winning the season-long championship battle at the end of the day: Davey Allison, Bill Elliott, Alan Kulwicki, Harry Gant, Kyle Petty, and Mark Martin.

The race also marked the end of one of NASCAR's longest running and historic careers, that of Richard Petty. The man deemed "The King" of NASCAR had competed on the stock car circuit for thirty-five years, compiling seven championship trophies, seven Daytona 500 victories and 200 total wins. Yet, the November 15, 1992, event at Atlanta Motor Speedway would be his final time behind the wheel in NASCAR competition.

On the same day that one of NASCAR's biggest stars was starting his final race, one of the sport's future stars was starting his first premier series event. After a disappointing campaign with Ricky Rudd and Ken Schrader in the 1992 season, Hendrick looked toward the future and took a gamble on a young hot shot from the open wheel ranks named Jeff Gordon.

Gordon was born in California, but his family moved to Indiana to pursue his racing dreams. Racing quarter midgets as a child, Gordon graduated to United States Auto Club (USAC) midgets and sprint cars, making a name for himself on ESPN's weekly *Saturday Night Thunder* racing program. In 1990, Gordon became the youngest midget champion in USAC history at the age of nineteen. One year later, he also became the youngest driver to win the USAC Silver Crown championship.

Through his successes in the open-wheel circuits among some of the sport's most experienced drivers, Gordon attracted offers from Toyota's stadium truck series, but he also had his eye on NASCAR and stock car racing. Gordon attended the Buck Baker Driving School and met independent driver-owner Hugh Connerty. After a few impressive laps, Connerty agreed to put Gordon in the car.

Gordon's first official NASCAR start came on October 20, 1990, in the AC-Delco 200 Busch (now Xfinity) Series race at Rockingham Speedway. As green a rookie as could be, Gordon put the Outback Steak House-sponsored Pontiac on the outside of the front row in his first qualifying attempt. It may have been Gordon's first venture into stock car racing, but he was already working with someone that would play a crucial role in his success over the next twenty years: Ray Evernham. On the 33rd lap Gordon struggled with a loose car and hit the outside wall exiting the second corner. The end results of his first race may have been disappointing, but his speed had once again turned some heads—especially that of Ford's Lee Morse.

A few weeks after the Rockingham race, Morse presented Gordon with an opportunity that was nearly impossible to refuse.

"I'm talking to Lee and he says, 'Would you be interested in the [Bill Davis] Carolina Ford Dealers car?' Excuse me? What? Every race I'd been to, that car flew," said Gordon.

Once partnered with Bill Davis, Gordon continued to work with Evernham and attract attention. Despite missing the season-opening race at Daytona, the 1991 season was a successful one as the pairing earned five top-five and ten top-ten finishes. However, the 1992 season would be a breakout year for the former open-wheel sensation.

Again partnered with Ford and Bill Davis, Gordon came out of the box strong, winning three poles in the his four races. On March 14, 1992, at Atlanta Motor Speedway, Gordon would start from the pole, lead 103 laps and beat Harry Gant by nearly four seconds to earn his first career NASCAR victory.

"It was a big one," Gordon said of that victory. "Not only was it my first win in NASCAR, but there were no slouches that I was racing against that day."

That season, Gordon would go on to earn a total of three wins, 10 top-fives, 15 top-tens and led a total of 1,060 laps. In addition, Gordon recorded 11 pole positions, breaking the mark set by Sam Ard.

However, the 1992 season was not without controversy. After landing his big break in NASCAR with Bill Davis, Gordon informed his team owner they would finish out the 1992 campaign together, but he had signed a contract to drive for Rick Hendrick in the NASCAR Cup Series the following year. Hendrick was impressed with his winning effort at Atlanta and called Gordon to court him to his multi-car team. Gordon agreed, so long as he could bring Evernham with him.

Not only was Gordon leaving the owner that gave him a break, but he was also leaving Ford for its biggest rival, Chevrolet.

The reception to the news was not well accepted by the Ford camp. Ford Motor Company's Michael Kranefuss told *National Speed Sport News*, "As far as I'm concerned, [Gordon] doesn't need to worry about Ford anymore. We put him in a Grand National, we invested a lot of money, made sure he got all of the right people, so why would he all of a sudden treat us like a piece of shit?"

Despite the controversy, Gordon would follow up his success at Atlanta Motor Speedway by sweeping the year's races at Charlotte Motor Speedway. Ignoring the bad press and hostility thrown his way, Gordon went to work proving the hype was real.

"I know there was a lot of controversy with me and Ford and Bill [Davis] moving forward not being with Bill," Gordon said years later. "I can easily say now that it was the right decision to go with Hendrick Motorsports and I can honestly say that as grateful as I was for Bill and Ford at that time, my life and my career would not be the same if Rick Hendrick hadn't of been calling me after that race."

As the 1992 season was coming to a close, Gordon and Hendrick were able to get backing from DuPont. To preview their 1993 efforts, the newly formed No. 24 team entered the season's final race. The goal for that first premier series start was simple, run as many laps as possible, learn as much as possible. and stay out of the way of Richard Petty and the championship contenders.

"I will certainly never forget that first race here. I will never forget that driver's meeting," Gordon said prior to his final race at Atlanta Motor Speedway in 2015. "That was an amazing driver's meeting. For it to be my first one in Cup and Richard's last one and the faces that were in there beyond just the drivers was pretty impressive. I know I have told this story many times; I still have that money clip that Richard handed out that day with my starting position. I wasn't that proud at the moment because we started twenty-first, but I will remember that forever."

Few could have predicted just how historic Gordon's first premier series start would be.

After the pomp and circumstance of the opening ceremonies, it did not take long for the drama to unfold on the racetrack. Pole-sitter Rick Mast got loose going into the first corner of the third lap and had the rear of his No. 1 Oldsmobile come around. The spin collected the second-fastest qualifier, Brett Bodine, and brought out the first caution of the day. The 500-mile race was already off to a raucous start, and championship contender Davey Allison suffered damage to the left rear of the car in the incident. The team went to work making repairs and avoided a potentially disastrous start to the day.

Over the course of the race, championship contenders would fall by the wayside. Mark Martin and Kyle Petty both suffered engine failures, ending their already slim hopes of winning the title. Harry Gant ran mid-pack for much of the day, never led a lap and was not a factor in the battle for the championship. That left three drivers—Allison, Elliott, and Kulwicki—to duke it out for the chance to write their names in the history book and hoist the champion trophy at the end of the day.

For Richard Petty, the goal was to go out there, stay out of trouble and finish the final race of his career without issue. However, when a wreck occurred in front of him on the 96th lap of the race, Petty's famous STP-sponsored No. 43 drove hard into the back of another car and caught on fire. Despite the heavy damage from both the contact and the fire, the Robbie Loomis–led team was able to somehow salvage the car, make as many repairs as possible and get him back out in time to finish the race on the track and not in the garage.

"Everybody always wants to go out in a blaze of glory, I just went out in a blaze," Petty said of the wreck. He would finish 35th in his final NASCAR start.

Gordon's day would not go much better. After a solid start, Gordon was running well and staying out of the way of the championship leaders. However, on Lap 164, Gordon's car broke loose and backed into the outside wall in the first corner.

"I was just pushing the car as hard as I could and pushed it too hard, and jumped in the gas a little bit too much in [turns] one and two, and around it went. I remember that part very vividly," Gordon said.

Like that, his first NASCAR premier series start was over. He would finish 31st.

On the racetrack the championship battle was as intense and close as ever. One of the favorites heading into the day, Allison had already suffered damage, but had rallied back—as he had done so many times during the 1992 season. Allison had overcome broken ribs, a broken wrist, severe crashes, and the death of his younger brother, Clifford, as well as his grandmother, but through it all the son of NASCAR Hall of Fame member Bobby Allison remained focused on winning the title.

Those hopes would come to an end with 77 laps to go when Ernie Irvan's car broke loose off the fourth corner and collected Allison's No. 28 Robert Yates–owned Ford. Allison made heavy contact with the inside wall and his day—along with his chances at the title—was over.

With Allison out of the picture, the battle for the championship was narrowed from a six-man race to a two-car battle between the powerhouse of Bill Elliott and his Junior Johnson–owned team and Alan Kulwicki's independent single–car operation. Earlier in his career, Kulwicki had turned down a high-dollar, high-profile offer from Johnson to compete for his own team, so it was only fitting the 1992 championship would come down to a battle of the favorite against the underdog.

The back-and-forth on the track was captivating to watch both in the stands and for those watching the live ESPN broadcast at home. One of the most calculating drivers to climb behind the wheel of a stock car, Kulwicki had figured out if he could lead the most laps he would win the championship, regardless of what Elliott did.

As green flag stops were underway in the closing laps of the race, Kulwicki stayed out one lap longer than Elliott and secured the bonus point for leading the most laps, 103 compared to Elliott's 102.

When the checkered flag flew on the final lap, Elliott and his Junior Johnson team had earned the win, his fifth of the year, but Kulwicki had been crowned the champion. The do-it-yourself, hard–nosed independent driver-owner had overcome the odds to earn his first NASCAR premier series championship.

Sadly, his reign as champion would not last long. Kulwicki was killed on April 1, 1993, when a plane he was in crashed in the Tennessee mountains en route to Bristol Motor Speedway. The plane had not been properly de-iced and lost power as it approached the final destination.

Just over three months after Kulwicki's tragic death, another major player in the 1992 championship battle would also lose his life. Davey Allison was attempting to land his helicopter at the infield of Talladega Superspeedway on July 12, 1993, when it crashed hard into the ground. Allison was transported to a local hospital where surgery was done to relieve pressure on his brain, but he would never regain consciousness, dying on July 13.

While Jeff Gordon's involvement in that historic event was just a small part of the overall story, his success in the years that followed made it clear the 1992 Hooters 500 was truly one of NASCAR's most important races. It also just happened to be Gordon's first.

NASCAR Invades Indianapolis

THERE ARE few things sacred in the racing world, but one of the highest on that list is Indianapolis Motor Speedway (IMS). The "Yard of Bricks" has been hosting automobile races since 1909 and has been the home to the Indianapolis 500 since 1911.

The idea of stock cars racing at Indianapolis, the Mecca of open-wheel racers, was sacrilege for nearly eighty years.

That all changed one afternoon in 1991, when four-time Indianapolis 500 winner and open-wheel ace A. J. Foyt took a stock car around the two-and-a-half-mile track following a television commercial shoot.

Following Foyt's laps in a stock car, NASCAR's Bill France Jr. and Indianapolis Motor Speedway CEO Tony George Jr. discussed the possibility of the famed speedway hosting a NASCAR event.

On June 22–23, 1992, NASCAR hosted a Goodyear tire test that brought NASCAR stock cars to Indianapolis Motor Speedway for the first time in history. That two-day test was seen by an estimated 30,000 spectators and included the likes of Davey Allison, Derrike Cope, Dale Earnhardt, Bill Elliott, Ernie Irvan, Mark Martin, Kyle Petty, Richard Petty, Ricky Rudd, Rusty Wallace, and Darrell Waltrip.

Rusty Wallace was the first driver officially on the track when it opened for the initial day of testing. Soon after, the rest of the drivers joined Wallace on the track to truly introduce the historic facility to stock car racing. The field ran a total of 341 laps on the first day of testing, plus an additional 251 on the second. Bill Elliott's Junior Johnson–owned Ford would record the fastest lap of the session at 168.767 miles per hour.

Legendary crew chief and former NASCAR executive Robin Pemberton worked at Sabco Racing at the time with driver Kyle Petty. Pemberton said that initial Goodyear tire test included a lot of "true racers."

"You only pull into a few of these places for the first time in your life—Daytona, Indianapolis, Darlington, all those places with a history." Pemberton said, "It was great. We had fun. It was low pressure. Teams worked together on a few different things. We staged a couple of short races. Going to Indy for that first time and being part of that tire test, I'm really glad to say I was part of that first group to go there."

Pemberton admitted that first trip to the Brickyard felt like the stock car world was stepping on the toes of the IndyCar world, but that quickly went away.

"I think if you're a real racer—real racers like all racing," he said. "You'll watch a good motocross race, a Formula One, CART, or IndyCar race and anything else. If you love racing, you love a good race. Indy was a great place to go."

On April 14, 1993, Tony George and Bill France Jr. announced the inaugural Brickyard 400 would be held on August 6, 1994. The history of both NASCAR and Indianapolis Motor Speedway would never be the same.

"One of the goals of most every race driver born in America was to race at Indianapolis," France Jr. said. "For somebody who elected to go and have a stock car career, that wasn't possible. Today it is."

Following the announcement, NASCAR held its first official test on August 16–17, 1993. A total of thirty-five teams took part in that official test, with NASCAR setting up mock races to see how stock cars would perform in race conditions at the track made famous by open-wheel cars.

When race week arrived, 86 cars attempted to qualify for the 43-car field. NASCAR altered the qualifying format to accommodate the massive field on the entry list.

H. B. Bailey made history when he became the first NASCAR stock car to make an official qualifying attempt at Indianapolis Motor Speedway, holding the unofficial pole for a brief moment. Dale Earnhardt would put his black No. 3 Chevrolet atop the leaderboard for much of qualifying, but it was Rick Mast who took the top spot with a lap of average speed of 172.414 miles per hour around the two-and-a-half-mile track.

Mast went down in the record books for earning the first NASCAR pole position at Indianapolis Motor Speedway, and he took home $50,000, a $40,000 van, and a $10,000 bonus for team owner Richard Jackson and the crew.

Not only did the inaugural stock car event at IMS attract big attention from racers, but it was also a huge draw for both media and fans.

Tickets to the historic event sold out within twelve hours when they went on sale in 1993, and the estimated crowd of 350,000 fans in the grandstands was the largest to see a NASCAR race until that point.

The race was also broadcast live on ABC, with open-wheel broadcaster Paul Page serving as host, while Bob Jenkins and former NASCAR champion Benny Parsons called the race from the booth.

In the opening segment of ABC's broadcast, Page introduced viewers to the history of Indianapolis and its brickyard foundation, its transition into the ownership of the Holman family, and how

Tony George and Bill France Jr. came together to put on the historic event before he turned his attention to the day's race.

After a quick run-through of the favorites to win—Dale Earnhardt, Rusty Wallace, Mark Martin, Kyle Petty, and Jeff Gordon—Page laid out the significance of the day's race, "By day's end, there will be new history. Four hundred miles from now, the Ray Harroun of this age will be known. A new tradition begins. NASCAR has come to the Brickyard."

Mary Fendrich Hulman, matriarch of the Hulman-George family, gave the command to fire engines, and for the first time in NASCAR history the roar of stock cars came to life at Indianapolis Motor Speedway.

Most in the garage and field expected Dale Earnhardt to quickly make his way past pole-sitter Rick Mast to lead the historic first lap—including Earnhardt's team owner Richard Childress.

Childress told ABC Sports's Dr. Jerry Punch prior to the race, "I know who money is laying on. Dale . . . We're going to run as hard as we can. I'm sure [Dale] has him a plan lined up to try and lead that first lap."

Starting on the outside, however, Earnhardt did not have position as the field raced into the first turn. Earnhardt's famous black No. 3 Chevrolet made contact with the right side of Mast's No. 1 Ford. The contact upset the handling of Earnhardt's car and he hit the outside wall exiting the fourth corner on the first lap.

Mast led the first official lap of NASCAR competition at Indianapolis Motor Speedway after holding off Earnhardt and Jeff Gordon. It was also the first time Mast had ever led the first lap of a race.

Earnhardt's determination to lead the first lap of the first race at Indianapolis had put him in the wall and sent him falling through the field in the opening laps, until a caution for debris on the fourth lap allowed the RCR team to work on his car.

As the race settled in, Gordon took command and gave very little leeway at the front of the field. A California native, Gordon's family moved to Indiana to pursue his racing career. The driver who grew up a few miles from Indianapolis Motor Speedway was in just his second season in NASCAR's premier series in 1994 but had earned his first career victory in the Coca-Cola 600 at Charlotte Motor Speedway earlier in the year.

Starting third, Gordon took the lead for the first time on the third lap of the race, getting around the pole-sitter, Mast. The rookie driver would lead the next 22 laps before Geoffrey Bodine got around Gordon for the top spot on Lap 25. As green flag pit stops cycled through, Gordon was back out front on Lap 48, staying on point for 31 of the next 33 laps until Bodine retook the lead.

With 70 laps to go in the race, Gordon caught the back bumper of Bodine's Ford to regain the lead. In NASCAR's first race at the famed track, the two drivers put on an amazing show for everyone watching in person, listening on the radio, or watching the live television broadcast. Lap after lap Gordon would get a run down the speedway's long straightaways and make a move on Bodine into the flat first corner.

When a hard crash between Dave Marcis and Mike Chase brought out the caution, Bodine and his brother, Brett, restarted the race at the front of the field on the 100th lap of the race.

Brett got a jump on the initial restart, but Geoffrey got a great run on his brother into the third corner and made contact with the left rear of Brett's car to take the lead. Going through the fourth corner, Brett paid his brother back with a bump of his own. That contact sent Geoffrey's No. 7 Ford spinning off the fourth corner in front of the field. As Brett drove on, Geoffrey's car slid back across the track as the rest of the field scrambled to avoid his damaged car. Clear of most the field, Bodine's car was slammed into by Dale Jarrett, heavily damaging both cars.

While Brett had raced his way to the front of the field, Geoffrey had been a figure at the front of the field for much of the day. The contact may have been slight in sight, but the brotherly feud was the culmination of a tense family situation that was months in the making.

"Brett spun me out," a disappointed Geoffrey told Jack Arute on ABC's coverage of the race. "We've had some family problems, some personal problems between us here lately, and he just unfortunately took it out on the racetrack with me. I never expected he'd do it. He's my brother, I still love him, but he spun me out."

According to Benny Parsons during ABC's broadcast of the race, the Bodine brothers had a run-in prior to a practice session at Talladega Superspeedway in May. The family feud carried on throughout the year and came to a head on the track in Indianapolis in front of millions of people.

Only twice in Indianapolis history had brothers led in the same race—Chevrolets and Unsers. Yet in the first stock car event at the famed track, Geoffrey, Brett, and their younger brother, Todd, were each credited for leading laps.

With Geoffrey Bodine behind the wall and out of contention, the race was really on for the second spot. Gordon's No. 24 rainbow-painted Chevrolet continued to dominate the race as Brett Bodine, Rusty Wallace, and Ernie Irvan all battled for second.

A late-race caution on Lap 131 brought the field to pit road and Wallace's crew sent him out front with the race lead, beating Gordon and Irvan off pit road with a 15.9-second stop.

With 28 laps to go, Gordon, Wallace, Irvan, Brett Bodine, and a host of other cars jockeyed for position side-by-side. After a stop under caution, Wallace was shuffled out of contention for the lead.

As had happened much of the day, the sophomore driver of the No. 24 Chevrolet was leading the field. This time, Irvan was filling Gordon's rearview mirror and putting pressure on for the lead in the closing laps.

Irvan and Gordon would swap the lead four times over the final 25 laps of the race, often racing side-by-side and within inches of one another. After a hard battle, Irvan led Gordon with 10 laps to go but the top six cars closed in on their bumpers.

With five laps to go, Irvan had a right-front tire go flat going into the first corner, and Gordon capitalized by diving under the No. 28 Ford to take the lead. As Gordon drove off with the lead over Brett Bodine, Bill Elliott, Rusty Wallace, and Dale Earnhardt, Irvan limped his car to pit road to the attention of his Larry McReynolds–led crew with no chance at victory.

"I was in a good spot. Even if I could get behind him, I was in a good spot because I could get him loose," Irvan told the ABC Sports broadcast. "Jeff did a good job. His car was great all day. We got ours right at the end. The good Lord was with us, he kept us from hitting the fence, but he couldn't keep us from running something over."

Irvan was the points leader heading into the Brickyard 400 but lost the lead to Dale Earnhardt after his ill-timed flat tire. Two weeks later, Irvan suffered life-threatening injuries during a hard crash in practice at Michigan on August 20, 1994. Irvan would return to racing and go to Victory Lane again but retired from the sport for good in 1999 after another big hit at Talladega Superspeedway in 1998.

As the capacity crowd rose to their feet, local racer, turned NASCAR star Jeff Gordon crossed the finish line first to take the checkered flag in the inaugural running of the Brickyard 400 at Indianapolis Motor Speedway at just twenty-three years old.

"In one lifetime one person doesn't get many chances to make an everlasting mark or set a record that no one will ever break," Gordon's winning crew chief Ray Evernham said. "When those things happen, it is the greatest payback that you will ever receive for the hard work and effort you've invested. The Brickyard 400 did those things for us."

Gordon would go on to win a total of five Brickyard 400 races at Indianapolis Motor Speedway and drove the pace car for the 99th running of the Indianapolis 500 in 2015. Looking back at that historic first race in 1994, the thing Gordon remembers most was the excitement of the fans.

"Most of the things that stand out to me was really about just the madness and craziness of how big that event was, how popular it was among fans, not just traditional NASCAR fans but new fans to the sport," Gordon said. "Even if you go back to the test that we had, the fans were just lined up on the fence around the garage area just wanting to see stock cars race at Indianapolis, and it was much of the same when it came to race day, just so many fans and you just couldn't walk anywhere without getting mobbed. That just showed you the impact and significance of that inaugural event."

Over the years, the NASCAR Cup Series event at Indianapolis began to lose its luster and was marred by an incident in 2008 in which tires were wearing so severely and quickly that NASCAR was forced to throw five competition cautions to allow teams the chance to check wear and change tires.

With attendance dwindling and viewership falling, this once crown jewel event on the NASCAR schedule needed a change.

In 2020, in front of an empty IMS due to COVID-19 restrictions, Kevin Harvick earned his second consecutive and third total Brickyard 400 victory, the final driver to win on the traditional oval in twenty-seven years.

Starting in 2021, NASCAR moved the Cup Series and Xfinity Series events from the oval to the infield road course, and a new chapter in NASCAR's relationship with IMS began.

On the weekend of August 13–15, 2021, NASCAR and INDYCAR shared the track for the first time in history, as both series competed on the IMS Grand Prix layout. The NASCAR Xfinity Series and NTT INDYCAR Series competed on Saturday, August 14, with the NASCAR Cup Series taking center stage on Sunday, August 15, in the Verizon 200 at the Brickyard.

"It's a fun course and it's an historic feel, which is something cool for our series and to be going down the front straightaway the opposite direction like in the old-school F-1 days like when they came here in early 2000," former NASCAR Cup Series champion Kurt Busch said following the weekend's practice session. "That's kind of the feel of everything. And then to have INDYCAR, Xfinity, and Cup…. everybody here, it's like the whole motorsports fraternity hanging out and it's really neat to see everybody's smiles. It's a lot of emotions today."

Much like the first NASCAR Cup Series event at IMS, the No. 24 Chevrolet was fast once again, this time with William Byron behind the wheel.

The First Use of Roof Flaps

RACING HAS always been an inherently dangerous sport, and one of the main areas of concern for NASCAR has been race cars flipping over and getting airborne. From the earliest days of the sport, cars have tumbled, rolled, and flown through the air.

By the 1993 season, things had gone too far.

During the season-opening Daytona 500, Rusty Wallace's No. 2 Pontiac was sent flying through the air and flipping twenty feet high after contact on the backstretch. Later that season, Wallace would go for another wild airborne ride after contact from Dale Earnhardt on the final lap of the Winston 500 at Talladega Superspeedway triggered the accident.

Wallace's wrecks were not the only major airborne incidents that season, however. Neil Bonnett also flipped in the tri-oval at Talladega, while Johnny Benson took a major tumble down the Michigan International Speedway backstretch in a Busch (now Xfinity) Series race that same year.

With so many violent airborne occurrences in the same season, NASCAR went to work figuring out a way to help keep cars on the ground.

The theory of the roof flap was simple. When a car was turned around backward, the two roof flaps would lift, allowing air to funnel into the cockpit of the car and creating more downforce. The resulting increase in downforce would, hypothetically, add more weight to the car and keep it on the ground.

Instrumental in the development of the roof flaps was team owner Jack Roush and his group of engineers. Roush does not take credit for the technological advancement, instead crediting others that worked alongside him.

"The credit goes to Chevrolet Wind Tunnel for defining really what the nature of the flap needed to be," said Roush. "They understood where the lift was coming from and how to stop it."

After a safety test at Darlington Raceway, in which most parties felt the results were a failure, a Chevrolet engineer suggested the idea of the dual roof flaps, but was unclear as to how to make them deploy during an incident. At that moment, Roush's technological mind went to work on how best to keep cars on the ground.

"I envisioned how to make them deploy on the roof, and I said, 'If you can give me a week, I can make you some prototypes that do just that. Just define for me exactly where you want them.' And he told me where he wanted them," said Roush.

The innovative team owner went back and worked with his engineers in Michigan with the exact specifications and of the flaps. Roush credits John Christian for being "most responsible" for the finished product.

"He took my ideas and converted them to hardware with springs and cables and carbon fiber parts and some steel parts for the hinges," he said. "The first set in racing lasted for only for about twelve, fifteen years."

Still, when the roof flaps were introduced in the 1994 season there were plenty of unknowns. Despite wind tunnel testing and engineering mockups, no one truly knew if the flaps would work as planned when needed most.

"A lot of guys called us up and said, 'Are you crazy? This is a lot of work, and a lot of trouble, and you don't even know if they work.' And obviously we couldn't ask somebody to drive their car 200 mph to spin out just so we could see if the roof flaps worked," recalled Gary Nelson, NASCAR's premier series director from 1992 to 2001. "So, we weren't sure. . . . We thought so but weren't sure."

The first true test of the new roof flap system came during the 1994 Daytona 500 qualifying race when Ritchie Petty, son of NASCAR Hall of Fame member Maurice Petty, spun off the fourth turn at Daytona International Speedway. When his car turned backward, the roof flaps immediately lifted, and the car was held securely on the ground without lifting into the air.

For Nelson, that single incident was enough to justify the work and expenses put into the new safety innovations.

"And me watching that, and seeing it happen, and knowing that the roof flaps did their job, gave me the biggest sense of accomplishment and pride," he said. "Just a good feeling that, hey, they're not gonna lynch me, number one, and number two, it really did work and it may do a lot to change the sport."

As the cars have continued to develop, so too have the roof flaps. From the older generation car to the Car of Tomorrow, to the Generation 6 car, and the completely revamped Generation 7 car, roof flaps have been a staple on each car since the 1994 season.

Just like the race cars, the roof flaps have evolved and adapted over the years, and NASCAR continues to work on ways to keep cars on the ground.

"We've made the roof flaps more effective by increasing their area and we've made them more high-tech with more advanced materials over the last fifteen to twenty years," said Roush.

However, cars continue to fly through the air, flip wildly, and be involved in spectacular wrecks.

Throughout each generation of car and advancements in safety, NASCAR continues to chase this ever-present danger in motorsports. Many inside the garage hoped the implementation of the Generation 7 chassis would help keep the race cars on the ground, but during its first season in use in 2022, there were two instances in which cars became airborne and rolled over—Harrison Burton in the Daytona 500 and Chris Buescher in the Coca-Cola 600. In both instances the roof flaps were deployed.

So while the roof flaps have been a great success in limiting the number of airborne incidents, they are certainly not an end-all to the problem that has plagued NASCAR since its earliest days. Technological work and investigations continue in this area and remains a focal point for the sanctioning body moving forward.

Let's Go Trucking

As THE American economy boomed in the mid-1990s, one of the side effects was an increase in automobile sales across the country. However, unlike any other time before, the top-selling vehicles in the nation were not family sedans, but instead pickup trucks. The Ford F-Series and the Chevrolet C/K pickup were the top-selling vehicles in 1994, leading some in NASCAR to take notice.

After team owner Jimmy Smith took an experimental race truck to Daytona International Speedway and ran some laps in February 1994, the conversations began to turn serious about the potential success of a truck racing series in NASCAR. Following that test run on Daytona's high banks, an exhibition truck race took place for the first time on July 30, 1994, at Mesa Marin Raceway in Bakersfield, California. Five trucks entered the 20-lap race, which was won by P.J. Jones. The race featured three lead changes among four different drivers.

When Craftsman Tools agreed to serve as the entitlement sponsor, the NASCAR SuperTruck Series presented by Craftsman was officially created, with the inaugural season to be run in 1995.

The new racing series developed interest right away. A number of premier series teams jumped at the opportunity to be involved, including team owners Richard Childress, Rick Hendrick, and Dale Earnhardt.

After a number of experimental races and test sessions were held, the first official NASCAR SuperTruck Series race took place on February 5, 1995, at Phoenix International Raceway. A total of 33 trucks entered the inaugural event, including some of the biggest NASCAR stars at the time: Terry Labonte, Ken Schrader, and Geoffrey Bodine.

Future truck series champion Ron Hornaday Jr. started the race from the pole position, with Schrader, Labonte, Johnny Benson, and Joe Bessey rounding out the top five. Taking the green flag on the historic event, Hornaday led the opening 23 laps of the race, before battling with Labonte for the top spot.

Mike Skinner, driving a No. 3 Goodwrench Chevrolet painted like Dale Earnhardt's car, took the lead for the first time on the 33rd lap, staying out front for the next 29 circuits. Labonte challenged for the lead once again on the 62nd lap of the race, remaining in the lead for the next 18 laps.

When the field bunched up for a restart with two laps to go, Skinner put the pressure on Labonte for the race lead. Racing tight on Labonte's bumper, Skinner made a move to the inside of Labonte for the lead as they raced through turns three and four, getting the advantage as they took the white flag for the final lap.

The two trucks raced side-by-side through the first two corners and down the backstretch, with Skinner finally pulling ahead of Labonte in the final corner. Blocking Labonte's crossover move and letting the right side of the truck slide out of the final corner, Skinner took the checkered flag to score the win in the inaugural truck series event.

After that inaugural event, the NASCAR SuperTruck Series competed in a total of twenty events for the 1995 season. Skinner would go on to win a season-high eight races to earn the first NASCAR truck series championship with team owner Richard

Childress. Joe Ruttman was second in the season-ending point standings with two victories, while Hornaday was third with six total wins.

The truck series races featured a ten-minute break in the middle of the race, to help reduce costs on the team and provide a level playing field for all teams, since some tracks did not have an official pit road. Those breaks allowed crews to make adjustments to the trucks and the drivers to be interviewed by the television broadcast, something that would help cultivate new NASCAR stars.

"We found out the TV coverage of the breaks produced an interesting side effect," said Dennis Huth, the NASCAR vice president in charge of the truck series in 1995. "It created some new stars. Where else could you have a full eight or ten minutes to talk with somebody in the middle of an event, like in a football locker room at halftime? Now, millions of households can see and hear what is done to the truck to make it competitive for the second half. You get a different philosophical view from the start of the race to the middle to the end. It's generated some new excitement."

After that inaugural season, NASCAR's chairman and CEO Bill France Jr. was pleased—and a little bit surprised—with the overall success and popularity of the truck series.

"The truck series is new and there's a lot of interest in trucks," France said in 1996. "New vehicle sales are heavily trucks and that shows interest. You take the makeup of that series and there's some good talent. There are a few Winston Cup drivers and we've got some new stars like Mike Skinner and Ron Hornaday being created. Some of these fellows have been racing a whole lot and they've had a chance for exposure in the truck series. But it wound up bigger than I thought it would."

Leading to more success than anyone could have planned for, NASCAR had big plans for the SuperTruck Series after that inaugural season.

"To say that NASCAR doesn't have some sort of master plan on the series . . . I'd be sitting here fibbing to you," Huth told *Stock Car Racing Magazine.*

The truck series would continue to grow and evolve in the years that followed the inaugural season of 1995. The term SuperTruck was dropped and the series became known simply as the NASCAR Craftsman Truck Series until Camping World took over the entitlement sponsorship role in 2009. With the 2023 season, Craftsman returned as the official partner of the series for the first time since 2008.

Over the first five years, Skinner, Hornaday, and Jack Sprague would emerge as the truck series' biggest stars, each taking turns as the champion. The halftime break was eventually eliminated from the series, and as it continued to grow both in popularity and in terms of competition, the small, short tracks gave way to intermediate tracks and superspeedways.

To kick off the 2000 season, the Craftsman Truck Series made its debut at Daytona International Speedway, doing so in dramatic fashion. The race featured two- and three-wide racing throughout the entire event but was marred by a horrific crash in the tri-oval.

Lyndon Amick made a bold move in the tri-oval to go under Kurt Busch and Rob Morgan, with Geoffrey Bodine just behind on the high side. When Busch moved up to keep from hitting Amick, he made contact with Morgan, sending his truck nose-first down the track and into Amick. The two trucks shot up the track and made contact with the left front of Bodine's truck, sending it airborne into the catchfence.

Bodine's truck disintegrated and bust into flames as it tore along the fencing. Tumbling wildly as the rest of the field piled in behind him, Bodine's mangled truck was hit multiple times after the initial contact. The engine was thrown three hundred feet down the track from where the truck finally came to rest, just a heap of bent roll cage and smoke. Bodine faced life-threatening injuries, but would recover.

The truck series would continue to evolve and change its role within NASCAR over the years, and became one of the most popular series in the sport. While there have been forty-one drivers to score wins in all three of NASCAR national touring series as of 2023, no driver has won the championship in all three.

One of the longest tenured drivers in the series, Hornaday leads all drivers with four Camping World Truck Series championships—1996, 1998, 2007 and 2009. In 2014, Matt Crafton made history by becoming the first driver to win back-to-back truck series championships. Crafton earned his third championship in 2019, tying Jack Sprague as the only other driver with three titles. Kyle Busch leads the series in all-time career victories with a total of sixty-three wins.

The truck series continues to feature some of the most competitive racing in NASCAR today. The series is made up of cagey veterans and young drivers still learning the ropes of the sport. What has turned into one of the most successful racing series in America can trace its beginnings back to a booming economy and a gamble by NASCAR to put trucks on the track in 1995.

Finally First: Dale Earnhardt Wins the Daytona 500

By the 1998 season, there were few things Dale Earnhardt had yet to accomplish in his illustrious career. The hard-nosed driver had cemented his place in NASCAR history by winning seventy NASCAR Cup Series races and seven premier series championships, tying the mark set by Richard Petty.

The glaring goose egg on Earnhardt's long list of achievements was a victory in NASCAR's biggest race, the Daytona 500. For nineteen years, Earnhardt made the trip to the two-and-a-half-mile Daytona International Speedway with the hopes of hoisting the Harley J. Earl Trophy. For nineteen years, Earnhardt walked away empty-handed and disappointed.

Heading into the 1998 edition of the race, Earnhardt had a total of thirty wins at Daytona, but never the "Great American Race." In addition, Earnhardt had led seventeen of his previous nineteen Daytona 500s, but never led the field to the checkered flag.

Earnhardt's tale of Daytona 500 heartbreak was the stuff of NASCAR legend. Aside from three engine failures in 1982, 1983, and 1985, Earnhardt's worst finish in the Daytona 500 prior to the 1997 season was 14th in 1986. However, poor luck and ill-timed misfortune seemed to always keep "The Intimidator" out of Victory Lane.

In perhaps his best opportunity to win the "Great American Race," Earnhardt's famous black No. 3 Chevrolet was leading a pack of single-file cars into the third corner on the final lap of the 1990 event. Racing hard down the backstretch, Earnhardt's car suddenly went to the high side, fell off the pace, and allowed Derrike Cope to slip underneath to score the win. Earnhardt had cut a tire on the final lap and could not keep the car in line through the final corners.

One year later, Earnhardt started fourth and quickly made his way to the lead on the second lap of the race. Even out front Earnhardt was not safe. Leading the pack in the opening laps, Earnhardt's car struck a seagull and damaged the right front nose of the car in the headlight area, right at the air inlet for the oil cooler.

Overcoming the damage from the seagull, Earnhardt remained a factor at the front of the field and was racing hard with Davey Allison for second with three laps to go. Chasing Ernie Irvan for the win, the rear of Earnhardt's car came around under Allison's No. 28 Ford as they exited the second corner and triggered a wreck at the front of the field. Earnhardt spun around and was clipped by Kyle Petty's car, heavily damaging the right front of the No. 3 Chevrolet. Despite the heavy damage to the car, Earnhardt limped home fifth.

In 1992, Earnhardt suffered slight damage during a massive wreck on the 93rd lap of the race that sent Rusty Wallace flipping nearly twenty feet in the air end-over-end. Once again with damage to the car, Earnhardt finished ninth.

Earnhardt was leading the way with two laps to go in the 1993 running of the Daytona 500 but could not keep a hard-charging Dale Jarrett at bay. Jarrett was able to get under Earnhardt in the third and fourth turns coming to the white flag for the final lap. Jarrett cleared Earnhardt on the final lap and was able to hold on for the win. Once again, Earnhardt had a Daytona 500 victory in sight only to have it slip away.

During the 1997 running of the Daytona 500, Earnhardt proved why many in the garage called him "One Tough Customer." Running in second to Bill Elliott in the closing laps, Earnhardt was doing all he could to keep Jeff Gordon, Dale Jarrett, and Ernie Irvan behind him.

With 12 laps to go in the race, Gordon made a move under Earnhardt off the second corner for the second spot. Earnhardt's car went high, bounced off the outside wall and came back across the track in front of Jarrett, who made contact with with the right rear bumper of Earnhardt's car. The contact immediately turned Earnhardt's car on its side as it headed for the outside wall. Rolling down the backstretch, Earnhardt's car finally came to a rest on the infield grass, damaged on nearly every corner. Earnhardt climbed from the car and walked into the ambulance but was not done just yet. When he saw all four tires were still inflated and pointed in the right direction, he got out of the ambulance to see if the engine would fire. When it did, Earnhardt climbed back in the car and finished the race five laps behind race winner Gordon.

As the historic 1998 season approached—NASCAR's 50th anniversary season—many wondered if Earnhardt would ever find Victory Lane in the Daytona 500. Earnhardt was in his second season with crew chief Larry McReynolds, but the pairing went winless during 1997.

"The thing that bugs me the most is how many times I've been so close and how many times I've lost it that way," Earnhardt said during ESPN2's 1998 Daytona 500 pre-race show. "I've had the race won by a mile, I've lost it by a half-mile, I've run out of gas, I've had a lot of scenarios happen to me. I've been passed on the last lap a time or two. I've been turned over a couple times."

When NASCAR roared into Daytona for Speedweeks, Earnhardt's No. 3 Chevrolet was among the strongest cars in the field, qualifying fourth in time trials. Earnhardt won his 125-mile qualifying race, leading each of the 50 laps to ensure a fourth starting spot in the weekend's main event.

The 1998 season also marked the first year Richard Childress Racing had expanded to two cars, bringing up Mike Skinner to drive the No. 31 Chevrolet for the entire season. For the first time in twenty years, Earnhardt would have a teammate on his side.

The weekend was not without issues for Earnhardt and his McReynolds-led team. During Saturday's final practice session ahead of the race, the car was down on power and running on seven of the eight cylinders. Not risking a failure in the race, McReynolds and the RCR team replaced the push rod in the engine and hoped the problem was solved.

For the first time in NASCAR history, a pair of brothers started the Daytona 500 from the front row. Bobby Labonte won the pole in his No. 18 Joe Gibbs Racing Pontiac, while his older brother Terry started second in his No. 5 Hendrick Motorsports Chevrolet.

On race day, weather threatened the start of the biggest race of the year. With a 70 percent chance of rain, overcast skies and strong winds would throw most drivers in the field a curveball throughout the 200-lap race.

NFL star quarterback Dan Marino waved the green flag to kick off the 50th anniversary season of NASCAR and the 40th running of the Daytona 500 in front of the largest crowd to attend the "Great American Race" until that point.

Earnhardt would take the lead for the first time on the 17th lap of the event. He would battle with Jeff Gordon, Rusty Wallace, Bobby Labonte, and Jeremy Mayfield throughout the race at the front of the field. By the halfway mark, Earnhardt was third behind Jeff Gordon and Rusty Wallace as the race remained caution-free.

With 61 laps to go, Earnhardt got around teammate Mike Skinner for the lead and never looked back. When a caution flew on with 26 laps to go, the field came to pit road for the final time of the day and Earnhardt's McReynolds-led pit crew was able to keep their driver out in front of the field.

Over the final 22 laps, Earnhardt was forced to hold off the Fords of Mayfield and Wallace, as well as the Chevrolet of Gordon and Pontiac of Bobby Labonte.

When John Andretti and Lake Speed wrecked off the second corner with two laps to go it was a race to the caution flag for the leaders. Approaching slower traffic off the final corner, Earnhardt was able to keep his No. 3 Richard Childress–owned Chevrolet out front as the yellow and white flags flew simultaneously.

Coming to the start-finish to take the caution CBS Sports' Mike Joy called the dramatic finish by saying, "Twenty years of trying, twenty years of frustration, Dale Earnhardt will come the caution flag to win the Daytona 500! Finally. The most anticipated moment in racing." Driving around under caution to complete the race, Earnhardt had finally kicked the monkey off his back and scored the victory that had eluded him for so many years.

"My eyes watered up in the race car," Earnhardt would say after the victory celebrations. "I don't think I really cried. My eyes just watered up on that lap to take the checkered. I knew I was going to win it then, no matter what. I knew I was going to win unless something happened to the car.

"I was driving slow down the backstretch and I said, 'I want to go fast. I don't want to go slow. I want to get back around there.' I took off, came back around, took the checkered, and really got excited."

After scoring the victory, Earnhardt's competitors congratulated him on the track during the cool-down lap, but few could have predicted what would happen when the "Man in Black" would bring his car to pit road.

Nearly every member of every crew walked onto pit road to offer their congratulations and shake Earnhardt's hand after the historic victory, one of the most popular in Daytona 500 history. After taking his time meeting all of the crew members, Earnhardt drove his No. 3 Chevrolet into the frontstretch grass, looping the car around three times to make an "E" with the tire marks.

"I sort of expected a few [competitors] to come out there, but not as many as there were," Earnhardt said of his reception on pit road. "All the guys came up congratulating me, all of them wanting to shake my hand or give me high-fives, thumbs-up. There was Michael Waltrip, Rusty . . . I had to go real slow or my arm would've gotten torn off."

Driving into Victory Lane, Earnhardt climbed from his car with a massive smile as he stood on the roof and threw both arms above his head in triumph. After so many years of frustrations and failures, he had finally won the Daytona 500.

"It was my time," he said. "That's all I can say. I've been passed here. I've run out of gas. I've been cut down with a tire. I've done it all. I wrote the book and this is the last chapter in this book. I'm going to start a new book next year. It's over with. Every which way you can lose it. I've lost it. Now I've won it and I don't care how I won it. We won it."

The dramatic victory in the season-opening race would be Earnhardt's lone win of the 1998 season and his only one in the Daytona 500.

"It was special because I knew what it meant to Dale's resume to win the Daytona 500," Childress recalled during RCR's 50th anniversary season. "He had won everything. We had won the Daytona 499 two or three times, but to get the 500 in 1998, I knew how special it was to him. To see all the crew members come out to pit road and stand and congratulate him—everyone in that garage knew how close we came but to finally pull it off was a big deal."

Three years after celebrating his Daytona 500 victory, Earnhardt would lose his life on the final lap of the 2001 running of the event. Earnhardt was busy keeping a hungry pack of cars at bay as the Dale Earnhardt Inc. cars of Michael Waltrip and Dale Earnhardt Jr. drove through the final corner to the checkered flag. Earnhardt would never make it back to the line, however, as his car bobbled in the fourth corner, hit the apron, and shot back up the track. Ken Schrader hit Earnhardt's right-side door as the No. 3 Chevrolet slammed hard into the concrete wall head-on.

While Waltrip was celebrating his first career NASCAR points-paying victory, Earnhardt was being transported to the local hospital, where he would be pronounced dead. The joy of the 1998 race was in direct contrast to the tragedy of Earnhardt's final race at Daytona. After his death on February 18, 2001, Daytona added a statue of Earnhardt outside the speedway. Earnhardt is holding the trophy from his Daytona 500 victory, arm raised in the air triumphantly and a smile on his face.

The tragedy of losing Earnhardt on the final lap of the Daytona 500 was something the entire sport had to overcome, but perhaps no one more than Richard Childress and his organization. As of 2023, there have been only two other Daytona 500s run on February 18, the date of Earnhardt's passing.

On February 18, 2007, Kevin Harvick honored Earnhardt by delivering RCR its second Daytona 500 in thrilling fashion. Harvick made an incredible charge from deep in the field in the closing laps to pass Mark Martin on the outside entering Turn 3 on the final lap. With the field crashing behind them, Harvick beat Martin to the line by 0.020 seconds.

In 2018, RCR marked the 20th anniversary of Earnhardt's Daytona 500 in grand style. Also run on February 18, Richard Childress's grandson, Austin Dillon, drove the iconic No. 3 Chevrolet to a dramatic victory in the "Great American Race."

With a huge run down the backstretch on the final lap, Dillon made contact with the back bumper of Aric Almirola's No. 10 Stewart-Haas Racing Ford and sent him spinning down the track. With no caution thrown, Dillon held off the rest of the field behind him to score his second career NASCAR Cup Series win and return the No. 3 RCR Chevrolet to Victory Lane in the Daytona 500 for the first time in twenty years.

"The most special part about this is my grandson, and twenty years later, we win the Daytona 500," Childress told the media following the 2018 win. "Dale winning that, I know how special the Daytona 500 was for his résumé. I was so happy for him to win that Daytona 500 because I knew how much it meant to him. And tonight, for Austin to come back, my own grandson, twenty years later and win the Daytona 500 is really—that's a special one. Both of them are special, but nothing is thicker than blood."

Childress knew his former driver and close friend would have been proud of the team's accomplishment all those years later.

"He would come over and grab Austin around the neck, put that old Dale Earnhardt hug on him, and tell him how proud he was of him," said Childress. "I knew Dale as good as anyone, and I know right now he's up there smiling down."

The First Fourth Generation Family: The Pettys

FROM THE very beginning, NASCAR has been a sport deeply rooted in family. The France family, the Flocks, the Bakers, the Allisons, the Earnhardts, Jarretts, Wallaces, and so on. Since the earliest days of the sport, fathers, sons, and brothers have shared their passion for racing, with the next generation often following in the footsteps of those laid before them.

One family has continually been a part of NASCAR since the first Strictly Stock race at Charlotte Speedway in 1949. When Lee Petty borrowed a Buick and rolled it during that first race, he started a trend that would continue for seven decades.

Lee's son, Richard, would soon join his father racing in NASCAR's Convertible and Strictly Stock divisions. In the 1980s, Richard's son, Kyle, would take up the family business with a style and flair all of his own. When Adam Petty made his first NASCAR Busch (now Xfinity) Series start at Gateway International Raceway on October 17, 1998, the Pettys became the first fourth-generation family in sports history.

The patriarch of the Petty bunch was Lee, who raised his family in rural Level Cross, North Carolina. While his neighbors made their living as farmers, the racing bug had bitten Petty, and after that first race in Charlotte in 1949 he was hooked.

Building his fleet of cars out of the family's house in Level Cross, Lee became one of NASCAR's earliest stars. He competed in six of the eight races held during the inaugural Strictly Stock division during the 1949 season, winning his first NASCAR race on October 2 at Heidelberg Speedway in Pittsburgh, Pennsylvania.

With little outside support, Lee's two sons, Richard and Maurice, helped their father in any way possible. Maurice would become one of NASCAR's master mechanics, earning him a spot in the Hall of Fame in 2014. Richard initially started out working alongside his brother, but he had his eyes set on driving. However, Lee would not allow Richard to get behind the wheel of a car until he was twenty-one years old.

When the time came, Richard Petty made his first NASCAR start on July 12, 1958, in a convertible division race at Columbia Speedway in South Carolina. Richard started the day 13th out of 25 cars and was able to climb to sixth when the checkered flag flew.

Six days later, Petty made his first NASCAR premier series start at the Canadian National Exposition Speedway in Toronto, Canada. Richard started seventh of 19 cars, but wrecked on the 55th lap of the race. Lee would win the race that day. Richard would go on to make nine starts during the 1958 campaign, with his first true season coming in 1959. He competed in 21 of the 44 events that season, recording six top-fives and nine top-tens en route to the Rookie of the Year Award.

Lee had 37 premier series victories and two championships under his belt when he was thrust into NASCAR history by winning the inaugural Daytona 500 in 1959 in a photo-finish with Johnny Beauchamp. The controversial race would take NASCAR officials sixty-one hours to determine the winner, but in the end it was Petty.

Lee's career behind the wheel came to an end during the 1961 season after he and Beauchamp were involved in a terrible crash on the final lap of the second 100-mile qualifying race at Daytona International Speedway. Beauchamp's car hooked Petty's right rear bumper, sending both cars soaring through the guardrail and down the large embankment of the fourth corner. Petty sustained life-threatening injuries in the crash, including a punctured lung. He would race just six more times over the next three years before officially retiring.

The leader of the Petty bunch, Lee ended his sixteen-year NASCAR career with 427 starts, 54 wins, 231 top-fives, 332 top-tens, and three championships in 1954, 1958, and 1959.

With Lee now sidelined due to his career-ending injury, the focus shifted to Richard's driving career. Richard's first career NASCAR premier series victory would come on February 28, 1960, at Charlotte Speedway, where Lee made his first start. Petty battled hard with Rex White for the lead, and thanks in part to a well-timed bump to White's bumper by Lee, Richard was able to hang on and earn his first victory.

Richard had come close to winning before, particularly on June 14, 1959, at Lakewood Speedway in Atlanta, Georgia. His No. 43 was initially flagged the winner of the 150-mile event, but a protest came from the second-place driver, Lee Petty. NASCAR officials determined the elder Petty was correct after an hour, moving Richard back to second place. It marked the first time a father and son combination would finish first and second in a NASCAR event.

Richard would go on to have plenty of success on his own, but all of it came with the Petty family on his side. Through his natural skills behind the wheel and a strong team behind him, Richard became "The King" of NASCAR and took the sport to new levels. Richard set the standard when it came to success on the track and fan interaction off the track. He would go on to set records that set the bar extremely high for those that followed: 1,184 starts, 200 wins, seven championships, seven Daytona 500 victories, 555 top-fives, 712 top-tens, and 123 poles.

When Petty retired from driving after the 1992 season, his son Kyle was already continuing the family tradition in a strong way.

A longtime figure at the racetrack, Kyle Petty had racing in his blood, but also loved to have fun in life. Growing up as a kid in Victory Lane and in the family shop in Level Cross, Kyle learned all he could from his father and grandfather, turning it into a career at the age of eighteen.

Kyle first climbed behind the wheel for an ARCA Racing Series event at Daytona International Speedway in 1979, winning his very first start in a stock car. From there, the expectations were high. Later that year Kyle would make his NASCAR debut, running his grandfather's No. 42, on August 5 at Talladega Superspeedway. Starting 19th, he would finish ninth in his first career start.

"Well, I wanted to race," Kyle recalled "My father said, 'If you're going to run Winston Cup cars, you might as well learn in Winston Cup cars.' That's what he believed. So at eighteen, I'm thrust into Winston Cup racing."

Kyle was much different from his father and grandfather. He was loose, liked to have fun, play his guitar, and joke around. He was often criticized for not focusing on the job at hand and for heading to Nashville to record a few country songs.

However, the third-generation driver would come into his own in due time. Kyle earned his first NASCAR premier series victory on February 23, 1986, at Richmond International Raceway. Driving for the Wood Brothers, Kyle slipped past Dale Earnhardt—who had dominated the race—with just three laps to go.

Kyle's best seasons came not with his family's race team, but with team owner Felix Sabates. During the 1992 and 1993 seasons, Kyle won three races and scored 18 top-fives and 32 top-tens en route to a fifth-place finish in the series standings both years. His NASCAR Cup Series career would continue until 2008, but during the 1999 season his focus once again shifted, this time to his eighteen-year-old son, Adam.

Much like his father, Adam Petty grew up in Victory Lanes and the infields of racetracks. When his famous grandfather or father would win, Adam was usually right there in the mix of things as a child.

Also like his great-grandfather, grandfather, and father, Adam found the racing bug and had a vision of continuing the Petty tradition into a new era of NASCAR. When he strapped into a Busch (now Xfinity) Series car in October 1998, history was made as the Pettys became the first four-generation family to compete in the same sport.

"There was never really a conversation where Adam said, 'I want to be a race car driver.' It was more, 'Hey, I like what you're doing, I want to be a part of this in some way, can I have a go-kart?'" Kyle told *BBC World Service*'s Sara Lentati in 2015. "Then one day you look over, and you're sitting in one car and he's sitting in a car beside you and you're racing."

Adam was young, bright, and beaming with personality. He often had the trademark Petty smile on his face and the talent of his grandfather behind the wheel. Honing his skills early on in the American Speed Association (ASA) circuit, Adam attempted the full 1999 Busch Series season with PE2, the team owned by his father. He earned his best finish at California Speedway in Fontana, California, bringing the car home fourth.

Through his first years behind the wheel, Adam worked very closely with his great-grandfather, Lee, to better hone his skills. Lee was not always easy on him, either.

"My dad puts no pressure on me and when I talk to my great-grandfather, he tells me I'm the worst driver in the world," Adam told *Inside NASCAR* during a 1998 interview.

The fourth-generation driver continued with the Busch Series in 2000, but also had his eyes set on NASCAR's premier series. On April 2, 2000, Adam made his NASCAR Cup Series debut at Texas Motor Speedway. While he and Kyle took part in the weekend's practice sessions together, Kyle failed to qualify for the event and was unable to race against his son. Adam would suffer an engine failure and finish his debut race in the 40th spot. Another father-son duo would enjoy the spotlight that day, as Dale Earnhardt Jr. earned his first victory, much to the delight of his father, Dale Sr.

As the 2000 race season continued, Adam's future was bright as the entire Petty family hoped he would carry the torch into the next era of NASCAR history. However, tragedy would strike the Petty family not once, but twice.

On April 5, just three days after Adam's first NASCAR Cup Series start, Lee Petty died at the age of eighty-six. From the earliest days, Lee was the Petty family's rock and guiding force. He laid the groundwork for his sons, Richard and Maurice, and set the stage for his grandson and great-grandson to follow his lead.

Just over a month after Lee passed away, on May 12, Adam was killed in a wreck during a Busch Series practice session at New Hampshire Motor Speedway. As Adam drove down the backstretch, the throttle on his car hung open and he hit the concrete wall hard, killing him instantly. He was just nineteen years old.

"I spent a lot of time away from home when I was seventeen and eighteen [years old], learning how to be a race driver," Adam once said. "I missed a lot of things that other kids my age were doing and I didn't get to spend a lot of time with my friends. But, man, it's worth it if I can make this happen."

Not only did the Petty family lose two members in a short amount of time, they also lost the future of their racing dynasty. With Kyle's career nearing the end, Adam was the hope and dream for a successful Petty driver to continue in NASCAR competition.

While the Petty family has enjoyed tremendous success in NASCAR over the years, they have also suffered their fair share of heartbreak.

To honor Adam's life, in 2004 the Petty family established Victory Junction, a camp for children with chronic medical conditions. In its nearly twenty years, Victory Junction has served over 115,000 experiences to campers in Adam's memory.

Both Richard and Kyle are still deeply involved in NASCAR to this day and can often be seen at the racetrack on the weekend.

Kyle has served as a television analyst and advocate of the sport in a variety of fashions since hanging up his helmet. Richard continued the Petty organization through the years with a variety of team names and partnerships, lastly with Maury Gallagher in the form of Petty GMS Motorsports. With Erik Jones behind the wheel of the No. 43, Petty GMS Motorsports won the Cook Out Southern 500 at Darlington Raceway.

Ahead of the 2023 season, fellow seven-time champion Jimmie Johnson invested into the organization, which was rebranded Legacy Motor Club, removing the Petty name for the first time since NASCAR's first premier series race in 1949.

"When Jimmie came in, it was going to be hard to be John-son Petty GMS," Petty said. "Jimmie's thinking further ahead with his crew and came up with a new name."

A new generation is following in the Petty family footsteps, despite not having the Petty last name. Thad Moffitt, the grand-son of Richard Petty, was born on September 12, 2000—four months after Adam's death in New Hampshire.

"I may not have the last name 'Petty,' but I'm as Petty as they come," said Moffit.

Moffitt made his way up the ranks as a young driver com-peting in late model series across the Carolinas, earning a championship in the Southeast Limited Late Model Series. The fourth-generation driver made a handful of starts in the ARCA Menards Series between 2017 and 2022, including his first start at Daytona International Speedway in 2019, qualifying second for that event. On February 18, 2022, Moffit made his NASCAR debut in the season-opening Camping World (now Craftsman) Truck Series race at Daytona, finishing 18th.

There is no doubt the legacy of the Petty name will carry long into NASCAR's future, even if it is no longer on the side of a race car. From the first Strictly Stock race in Charlotte in 1949 to nearly every other iconic moment in the sport's history, the Petty family have left a lasting mark on the sport's seventy-five-year history.

NASCAR Gets SAFER

FOR MUCH of NASCAR's history, danger was lurking around every corner for competitors. In the infancy stages of the sport, racers drove on makeshift dirt tracks scattered around the country, often with chicken-wire fences or wood boards keeping the cars on the track.

As the sport evolved and graduated to the paved tracks, guardrails and concrete retaining walls were erected around the tracks. Much more effective than their predecessors, these retaining walls offered their own host of dangers and potential issues.

In 1958, Eddie Pagan ripped out roughly one hundred feet of the guardrail at Darlington Raceway before leaving the track and flipping end-over-end. In the May 14, 1960, running of the Rebel 300 at Darlington, Johnny Allen's Chevrolet flew over the guardrail, out of the track, and into the scoring stand, collapsing the left side and trapping forty-eight people. Allen and those on the scoring stand survived the incident.

In a 1961 Daytona 500 qualifying event, Lee Petty and Johnny Beauchamp tangled in the fourth turn, flying through the guardrail and out of the track, ending the careers of both of the earliest stars of the sport.

As time went on, NASCAR moved away from the guardrails and implemented concrete retaining walls at nearly all their facilities. While concrete did a much better job of keeping cars inside the track, they did little to reduce injuries during a big hit.

During the 1990s and 2000s, as speeds increased across the board in NASCAR, the dangers of concrete walls became very clear.

Veteran driver Neil Bonnett was killed in a wreck while testing at Daytona International Speedway in 1994. Bonnett lost control of the car and it veered nearly head-on into the concrete wall.

Unfortunately, Bonnett was not the only driver to lose his life after a hard impact with a concrete retaining wall in the 1990s. Drivers such as Clifford Allison, Rodney Orr, and John Nemechek all lost their lives after hitting a concrete retaining wall.

However, from May 2000 until October 2001, five drivers were killed in accidents in which they hit unprotected concrete walls.

Fourth-generation racer Adam Petty died on May 12, 2000, at New Hampshire Motor Speedway after his throttle stuck and his car hit the concrete wall at nearly full speed. Just under two months later, Kenny Irwin Jr. was killed at the same track, in the same turn and in the same fashion.

In October 2000, Tony Roper lost his life in a NASCAR Craftsman Truck Series event at Texas Motor Speedway.

Yet it took the death of NASCAR's biggest star, Dale Earnhardt, to truly cause a wakeup call to the sanctioning body, drivers, team owners, and racing facilities.

Earnhardt's death came on the grandest of stages, the Daytona 500. The Intimidator was running third behind the cars he owned driven by longtime friend Michael Waltrip and his son, Dale Earnhardt Jr.

On the final lap of the race Earnhardt was blocking a hungry pack of cars to allow the two Dale Earnhardt Inc. cars to win the race. The black No. 3 Chevrolet bobbled in the corner after slight contact with Sterling Marlin, hit the apron of the track, shot up the banking where he collected Ken Schrader, and hit the concrete wall head-on at nearly full speed.

A number of factors went into each of these racing deaths. However, in the wake of some of the sport's darkest and most tragic days, NASCAR and the racing facilities began to make changes, but they take time, money, and thorough planning. Despite safety innovations such as full-face helmets and the Head and Neck Support (HANS) devices, racing remained a dangerous sport in the years that followed Earnhardt's death.

Almost nine months after that tragic day in Daytona, another Earnhardt was involved in a frightening crash that cost the life of Blaise Alexander. Racing for the lead with less than five laps to go during the October 4, 2001 ARCA Racing Series event at Charlotte Motor Speedway, Kerry Earnhardt, Dale's oldest son, and Alexander were racing for the lead as they caught slower traffic in Turn 4.

Alexander had a great run on the bottom and took the lead from Earnhardt off the corner. However, the two made contact in the tri-oval and Alexander's car was sent hard into the outside wall nearly head on before ricocheting back into the right rear tire of Earnhardt's car, flipping it upside on the roof and sliding down the frontstretch in a shower of sparks.

After a few scary seconds, Kerry climbed out from under the car and ran to the infield wall. However, there was no movement from Alexander's heavily damaged car. The hard hit into the concrete retaining wall killed the twenty-five-year-old almost instantly.

On May 2, 2003, during NASCAR Cup Series practice at Richmond International Raceway, Jerry Nadeau's car spun backward into the first corner and slammed the concrete retaining wall hard on the driver side of the car. Safety workers had to cut Nadeau from the wrecked car, but he would survive. The massive hit caused head, lung, and rib injuries to Nadeau, and he would never race again.

Continuing to react and develop new ways of protecting competitors, NASCAR got a bit of help from their open-wheel compatriots in 2002 with the introduction of Steel and Foam Energy Reduction (SAFER) barriers.

Developed in partnership with IndyCar CEO Tony George, along with the University of Nebraska-Lincoln, SAFER barriers were first introduced at Indianapolis Motor Speedway in 2002 for the 86th running of the Indianapolis 500.

The steel wall is separated from the concrete wall by energy-absorbing foam, which acts as a cushion when a car comes in contact with the wall. While damage to the car and the potential for injury still remained, the SAFER barrier was leaps and bounds above the traditional concrete wall.

When the NASCAR Cup Series hit the track in August 2002 for the annual Brickyard 400, it marked the first time the series would race at a facility surrounded by the new softer walls. Other facilities followed suit, with New Hampshire Motor Speedway—the site of both Irwin and Petty's fatal crashes—the next facility to install them.

Eventually, nearly every track at the NASCAR circuit would have some section of the track—often the corners—covered with the SAFER barriers. Since Dale Earnhardt's death in the 2001 Daytona 500, there have been no fatal accidents in NASCAR.

"Before Dale's death, we were kind of arrogant in thinking that we were the ones who built, wrecked, and repaired the cars, so we were the only experts," said former NASCAR vice president of research Gary Nelson. "The idea was to go outside the garage and say, 'Maybe there's someone else who can help us.'"

As tracks increased the area of walls covered with SAFER barriers, drivers continued to find areas where there was still bare concrete. Sterling Marlin was in championship contention until he hit the outside concrete wall at Kansas Speedway on September 29, 2002, and suffered a fractured vertebrae.

If any driver had a knack for finding unprotected concrete walls, it was Jeff Gordon.

During the March 2, 2008, Cup Series race at Las Vegas Motor Speedway, Gordon and Matt Kenseth wrecked off Turn 2. Gordon's car slid head-on into an angled non-SAFER barrier wall along the inside of the backstretch.

Gordon took another hard hit on a concrete wall during the April 30, 2011, race at Richmond International Raceway. After contact with Kenseth, his car spun down the track and slammed into an angled concrete wall on the driver side.

Luckily, Gordon would walk away from the incidents, but has cited both as contributing factors to the back pain that plagued the later years of his career.

In 2015, the topic of SAFER barriers was once again thrust to the forefront of conversation in the NASCAR garage. During the season-opening NASCAR Xfinity Series race at Daytona International Speedway on February 21, Kyle Busch was involved in a multi-car wreck exiting the tri-oval late in the race. Busch's car slid down the long apron of the track, shedding little speed as it fast approached the inside wall. Slowing down to roughly 90 miles per hour, the car struck the non-protected concrete wall.

Busch was able to climb from the heavily damaged car, but immediately went to the ground and was attended to by track safety workers. He would suffer a compound fracture of the right leg and a mid-foot fracture of the left foot, both of which required surgery.

The night of Busch's contact with the concrete wall, Daytona track president Joie Chitwood III and NASCAR Executive Vice President and Chief Racing Development Officer Steve O'Donnell addressed the issue of Busch hitting a non-SAFER barrier wall and promised changes would follow.

"The Daytona International Speedway did not live up to its responsibility today," Chitwood said that night. "We should have had a SAFER barrier there today, we did not. We're going to fix that. We're going to fix that right now."

That evening track workers put tire packs in place along the concrete wall Busch hit, causing his injuries.

O'Donnell promised NASCAR would seriously investigate the issue and make adjustments to improve the overall usage of SAFER barriers.

"I think we all know that racing is an inherently danger-ous sport, but our priority is safety and we'll continue to put things in place that make this sport as safe as possible," he said.

One week after Busch's vicious wreck at Daytona, Jeff Gordon was involved in a late-race incident on March 1 at Atlanta Motor Speedway. Gordon's No. 24 Hendrick Motorsports Chevrolet slid hard into the inside backstretch wall, which was not covered by a SAFER barrier.

Despite Atlanta adding tire packs along the inside wall in the corners, Gordon once again found a non-protected concrete wall. He was not happy with it, either.

"I am very frustrated with the fact there are no SAFER bar-riers down there," Gordon said after the incident. "I knew it was a hard hit. I was like, 'Man I can't believe [it].' . . . I didn't expect it to be that hard. Then I got out and I looked and I saw, 'Oh, wow, big surprise I found the one wall here on the back straightaway that doesn't have a SAFER barrier.'

"I don't think we can say any more after Kyle's incident at Daytona," Gordon added, reminding everyone of Busch's injury the week before. "Everybody knows we have to do something and it should have been done a long time ago. All we can do now is hope they do it as fast as they possibly can."

The incidents sparked another wave of safety innovations, with NASCAR, International Speedway Corporation, Speedway Motorsports Inc., Dover International Speedway, and Pocono Raceway announcing plans to survey each track and add SAFER barriers where needed.

At the time, every facility with the exception of Sonoma Raceway utilized SAFER barriers at least along some sections of the wall.

"Thankfully there's a lot of people paying attention to it and we can get this addressed," Hendrick Motorsports' Jimmie Johnson said after Gordon's incident at Atlanta. "It's unfortunate we're so many years removed from the inception of the SAFER barrier. I think we're on the right track and have it everywhere it needs to be."

The former champion did not let his teammate and good friend Gordon off the hook, however.

"We just need to send Jeff out on the track and find the places to put them. That guy will find the openings," he joked.

As the 2015 season continued on, NASCAR and tracks across the board announced plans to cover more walls with SAFER barriers as the sport continued to chase the ever-moving target that is safety in motorsports.

Today, SAFER barriers are an integral part of track safety in NASCAR, but it all started with a little help from Indianapolis Motor Speedway.

Chasing the Championship

WHEN BRIAN France was named NASCAR's Chairman and CEO on September 13, 2003, he would run the company much differently than his grandfather and father before him. Unlike Bill and Bill Jr., he was a quiet executive, who was not a prominent figure week-to-week at the racetrack, but hard at work behind the scenes. However, Brian had a vision for the future of NASCAR, and he went to work implementing his plan right away.

The changes would be big and bold. Racing back to the caution flag was eliminated just five days after he assumed his new role. After the 2004 season, team owners were limited to a four-car operation in the NASCAR Cup Series. In addition, the top thirty-five cars in the owner standings would be locked into each race to ensure the top drivers didn't miss the event.

Yet the biggest change of all came in January 2004, when Brian France announced an overhaul to the championship point system that had been in place since 1975. During the preseason media tour, France announced the new championship format would be called "The Chase for the NASCAR Nextel Cup," a playoff system for stock car's premier division.

Looking to expand the sport to new markets, untapped potential sponsors and the casual fan, France's new championship format moved NASCAR closer to the traditional stick-and-ball sports such as football and baseball. The decision also followed the 2003 season, which saw Matt Kenseth earn his first NASCAR Cup Series title after winning only one race all season long. Kenseth finished 90 points ahead of Tony Stewart and wrapped up the title with one race left in the year. He was only the fourth driver to win a NASCAR premier series championship with just one win.

Under the format, the top ten drivers in the point standings after the first twenty-six races would compete for the championship over the final ten races of the season. During the regular season, the point system would remain largely unchanged, with a victory earning 180 points as opposed to 175.

However, after the first twenty-six races, the top ten drivers in the standings would have the points adjusted. The top driver was given 5,050 points, with the point increments going down by five for each driver. After the reset, the tenth driver was only 45 points out of the lead. Once the Chase officially kicked off, drivers would continue to accumulate points as they had in the previous twenty-six races, with the champion being the driver with the most points.

The Chase would kick off at New Hampshire Motor Speedway, making stops at Dover Motor Speedway, Talladega Superspeedway, Kansas Speedway, Charlotte Motor Speedway, Martinsville Speedway, Atlanta Motor Speedway, Phoenix International Raceway, Darlington Raceway, and ending with the championship weekend at Homestead-Miami Speedway.

Despite some concerns from fans, competitors, and media alike, NASCAR maintained the Chase format was the right move for the sport and created a fair and level playing field for all competitors.

"The Chase for the NASCAR Nextel Cup will provide a better opportunity for more drivers to win the championship, creating excitement and drama throughout the entire season," said NASCAR President Mike Helton. "In addition, the Chase will showcase our drivers' talents, increasing the value for all teams and their sponsors.

"This new approach to determining our champion has both the drivers and the fans in mind."

Before deciding on the new Chase format, the third-generation France talked it over with his father.

"My dad thought it was real radical," Brian said. "I remember when I first told him about the idea, he told me to let him sleep on it, and we'd talk about it the next day. He acknowledged it was radical, let's say, in his own way. But I think he knew it was something I wanted to do. He was trying to be supportive of my leadership, and in the end he was always supportive—not only of me but of anybody who was trying to lead and trying to have responsibility and expectations attached to them. He knew you had to give people some room to fail or to succeed. He was pretty good about that."

While the focus would be on the final ten events of the year, the season-long fight to make the Chase would create unexpected drama for the twenty-sixth race of the season at Richmond International Raceway.

With a chance at competing for the championship on the line, Jeremy Mayfield started his Ray Evernham–owned Dodge from the seventh spot, led five times for a total of 151 laps to earn the victory and advance into the group of ten Chase drivers.

Mayfield was in a must-win situation going into the Richmond race. After finishing second one week earlier at California Speedway, Mayfield's teammate Kasey Kahne was ninth in the point standings. Mayfield entered the race 14th in the standings behind Kahne, Mark Martin, Jamie McMurray, Bobby Labonte, and Dale Jarrett. While Mayfield led the most laps and won the race, he received a bit of luck when Kahne was spun out by Dale Earnhardt Jr. on the 207th lap of the race. Kahne would finish 24th, fall to 12th in the standings, and miss the inaugural Chase for the NASCAR Nextel Cup.

Following the final "regular season" race at Richmond, the ten Chase drivers had their points reset according to the new rules. Jeff Gordon maintained his point lead, followed by Jimmie Johnson, Dale Earnhardt Jr., Tony Stewart, Matt Kenseth, Elliott Sadler, Kurt Busch, Martin, Mayfield, and Ryan Newman.

When the Chase kicked off at New Hampshire Motor Speedway, Kurt Busch and his Jack Roush–owned team came out swinging. Busch led three times for a race-high 155 laps to earn the victory over fellow Chase drivers Kenseth and Earnhardt.

Earnhardt and Busch were tied for the lead in the Chase standings, with Gordon and Kenseth in hot pursuit.

During the first race, three championship contenders took a big hit right off the bat. Stewart and Mayfield were involved in an accident started by Greg Biffle and Robby Gordon, while Newman suffered an engine failure and was forced behind the wall.

Newman would make up for it the following week at Dover Motor Speedway, leading 325 of the 400 laps to earn the victory. However, the win did little to move him up in the championship standings, which Gordon took command of by one point over Busch.

Earnhardt struck next, winning at Talladega Superspeedway with a charge from 11th on the final restart of the race with five laps to go. When asked by NBC's Matt Yocum about the win in Victory Lane, an overly excited Earnhardt said, "It don't mean shit right now. Daddy's won here ten times." The next day NASCAR fined Earnhardt $10,000 and docked him 25 championship points, bumping him from the lead and giving Busch the advantage.

From there, Busch never relinquished the lead in the Chase standings. However, things were tightening up as the season finale approached. Going into the ninth race of the Chase at Darlington Raceway, the top four drivers in the standings—Busch, Gordon, Earnhardt, and Johnson—were separated by just 48 points.

With two races left in the season, Johnson jumped from fourth to second in the Chase standings by edging Mark Martin for the victory in the Southern 500. Entering the race 48 points behind Busch in the Chase standings, Johnson went into the final race of the season just 18 points out of the lead.

Heading to Homestead with the championship on the line, Busch put his Jimmy Fennig-prepared Ford on the pole for the 400 mile race. Johnson, on the other hand, struggled in qualifying and started the final race of the year from the 39th spot.

Once the race got underway, the action was hot and heavy on the track. The race was slowed fourteen times for cautions and featured fourteen lead changes among seven different drivers. Busch would lead the first four laps of the race but would have multiple issues that put his championship hopes in danger.

In one of the most dramatic moments of the season, Busch slowed while running second on the 93rd lap with a bad vibration. As he drove onto pit road from the access road, the right front tire broke away from the car and rolled down the front straightaway. Headed straight for the barriers protecting the end of pit wall, Busch took evasive action, kept the car on pit road, and continued on to his pit stall on three wheels. The team was able to change tires and send their driver back into the race.

The team had another setback under the fifth caution of the day. A miscommunication from crew chief Jimmy Fennig turned a two-tire pit stop into a four-tire pit stop, costing Busch valuable time and track position. With drama unfolding for Busch on pit road, the championship battle between Busch, Gordon, and Johnson was back-and-forth.

When Ryan Newman blew a tire while leading, it set up a restart with just two laps to go. Lining up fourth for the restart, Gordon had a shot at the title if he could score the victory. When the green flag flew, Gordon got a great jump but could not capitalize as Greg Biffle moved around Tony Stewart to take the lead. As his teammate took the checkered flag, Kurt Busch brought his Roush-owned Ford home in the fifth spot behind Biffle, Gordon, Johnson, and Stewart.

Thanks to his efforts to charge back from adversity, Busch was able to edge Johnson by just eight points in the Chase standings. With consistency and a little bit of racing luck on his side, Busch earned his first NASCAR premier series title, delivered the first championship for team owner Jack Roush, and became the first winner under the Chase format. The new format had created the closest championship since Alan Kulwicki beat Bill Elliott for the 1992 title by just ten points and was the smallest margin of victory in the fifty-six year history of the series to that point.

The Chase would undergo further changes over the years, each with an eye toward placing an increased emphasis on winning and tighter championship battles. The Chase field was expanded to twelve drivers, and then again to sixteen, the point system used in the Chase was altered, as were the tracks that made up the final ten races. For the 2014 season, however, Brian France would once again overhaul the championship format entirely.

The First Winner-Take-All Championship Race

RACING FOR the NASCAR premier series championship took on an entirely different look starting with the 2014 season.

During the annual Charlotte Motor Speedway preseason media tour, NASCAR Chairman and Chief Executive Officer Brian France announced a major overhaul to the Chase for the Sprint Cup.

The Chase field was expanded from twelve to sixteen drivers, and under the new rules a win in the regular season essentially locked a team into the Chase. The sixteen drivers would include the regular season winners and the highest-ranked drivers in the top-thirty in points without a win.

Those drivers would compete for a shot to race for the title in the season's final race at Homestead-Miami Speedway. The Chase races were divided into three rounds, three races each. After three races, the lowest four drivers were eliminated from title contention. A win in any round of the Chase automatically qualified a driver for the next round.

In the end, four drivers would compete for the Sprint Cup championship at Homestead-Miami Speedway in a winner-take-all championship race. The four title contenders would score no points or bonus points for laps led. Instead, the highest finishing of the four contenders would earn the championship.

"It's going to elevate racing," France said while introducing the format in January 2014. "It's going to make winning the most important thing by a wide margin. It's going to change the strategies . . . This will be different. Everything is focused around winning, and that is exactly what our fans want."

This marked the fourth change to the championship format in eleven years, after the third-generation France introduced the Chase format for the 2004 season. This change, however, was geared to put an emphasis on winning and create the "Game Seven" moment France had wanted for many years.

"We have millions of fans and we have some very loud and passionate fans, especially when we change anything. We understand that," France said. "The vast majority of the fans that we communicated with, and I think we're the best in sports in staying close to our fan base, really love this. They love it because they really don't like points racing. At the end of the day, although consistency is important in our sport, and it remains important, it's just less important, so they like that. They understand winner-take-all formats, and they understand being the best down the stretch."

While skeptics remained, the inaugural season under the new Chase turned out to be thrilling, drama-filled, and must-see.

The inaugural Chase field under this format included winners Brad Keselowski, Jeff Gordon, Dale Earnhardt Jr., Jimmie Johnson, Joey Logano, Kevin Harvick, Carl Edwards, Kyle Busch, Denny Hamlin, Kurt Busch, Kasey Kahne, Aric Almirola and A.J. Allmendinger. Matt Kenseth, Ryan Newman, and Greg Biffle made the Chase field without a victory on the season.

As the Chase field was whittled down, the drama of each race intensified.

The elimination format forced drivers and teams to risk it all to advance. That led to post-race confrontations between

2012 champion Brad Keselowski and former champions Matt Kenseth and Jeff Gordon.

After the Chase race at Charlotte Motor Speedway, Kenseth ran down Keselowski between the haulers, sparking a shoving match between drivers and teams in the tight confines of the garage. Kenseth's Joe Gibbs Racing teammate Denny Hamlin had also tried to confront Keselowski after the race, but was restrained by crew member.

Watching the Kenseth-Keselowski confrontation unfold on Charlotte's big screen television looming over the backstretch, Dale Earnhardt Jr. smiled told the group of gathered media members, "The damn Chase is working."

Keselowski overcame the adversity and animosity against him to earn a must-win victory the following week at Talladega Superspeedway to advance into the next round of the Chase.

Three weeks after his run-in with Kenseth at Charlotte Motor Speedway, Keselowski was involved in another Chase-driven post-race scrum, this time with Jeff Gordon at Texas Motor Speedway.

With a win and a spot in the championship four on the line, Keselowski made a bold three-wide move on Gordon for the lead late in the race. The two made contact, Gordon's left rear tire was cut, and his chances at the title slipped away.

After the race, Gordon and his crew members confronted Keselowski on pit road. The two drivers exchanged heated words, and after a Kevin Harvick shoved Keselowski back into the scrum, fists began to fly. Both Gordon and Keselowski walked away bloodied, and crew members from both teams exchanged blows.

In the end, the championship four was determined at the 35th race of the season at Phoenix Raceway. In a deep hole after a 33rd-place finish at Martinsville Speedway, Kevin Harvick showed up to Phoenix and dominated, leading 264 of the

312 laps to earn the victory and punch his ticket to the championship round.

Denny Hamlin and Joey Logano also advanced to the title race at Homestead-Miami Speedway, while Ryan Newman used a daring move on the final lap of the Phoenix race to bump his way past Kyle Larson and gain the crucial spot to advance. Newman was the only driver among the four without a victory on the year.

Prior to the 2014 season-finale at Homestead-Miami Speedway, there had never been a winner-take-all championship race.

The NASCAR premier series title had been decided in the final race on numerous occasions, such as when Alan Kulwicki won the title over Bill Elliott in 1992 in the final race, and Tony Stewart tied Carl Edwards in 2011, and earned the title in a tiebreaker. But never before had the championship been determined by a no-points, highest-position-wins race.

Four of the best drivers all season long, Kevin Harvick, Denny Hamlin, Joey Logano, and Ryan Newman put on a show for the ages contending for the championship on November 16, 2014.

From the drop of the green flag, Harvick, Hamlin, Logano, and Newman found their way to the front of the field and stayed there. The four drivers raced in the top five for the majority of the race, keeping tabs on each other throughout the entire event.

The title came down to the final round of pit stops. Harvick and Newman opted for four tires, while Hamlin took just two. Logano's team took four tires as well, but the jack broke on the left side of the car, slowing the stop and dropping him outside the top 20 for the final restart of the race.

As the green flag run went on, Hamlin's car struggled on older tires and Logano was able to make little forward progress.

On the other hand, the cars of Harvick and Newman came to life and raced to the front of the field.

While neither was required to win the race to earn the championship, the pair raced to the front of the field and battled for the win over the closing laps. Harvick took the lead from Hamlin with eight laps to go and was able to keep Newman at bay over the final laps of the race.

With his fifth victory of the 2014 season, Harvick earned his first NASCAR premier series championship and the first ever winner-take-all championship race.

"In the end, it turned out you had to go for broke just to be competitive," Harvick said of the winner-take-all race at Homestead. "I think that's really what this format has turned every week into over the last ten weeks is if you want to win, you've got to—if you want to win the championship, you're going to have to figure out how to win races, and in the end, that's what it came down to was winning the race, and obviously a gutsy call and four tires on the pit box. In the end you had to win the race to win the championship, and it all worked out."

Five-for-Five: Jimmie Johnson's Five Straight Championships

PERHAPS NO driver adapted to the Chase for the NASCAR Cup format than Hendrick Motorsports's Jimmie Johnson. The California native seemingly burst onto the NASCAR scene out of nowhere, but etched his name as one of the best in the history of the sport along the way.

Discovered in the early 2000s at a test session by Jeff Gordon, Johnson was selected to drive at Hendrick Motorsports for a new fourth team with longtime Hendrick employee Chad Knaus serving as his crew chief. Prior to his premier series debut at Charlotte Motor Speedway in October 2001, Johnson was known primarily for a dramatic crash in a Busch Series race at Watkins Glen International where the brakes failed at the end of the straightaway and his car flew headfirst into the styrofoam blocks protecting the tire barriers. With most fearing the worst, Johnson stood atop his car, arms extended high above his head as if he had just won the race, simply thankful to have survived.

Ten years after that frightening crash, Johnson would once again get out of his car and extend his arms into the sky, this time hoisting his fifth consecutive NASCAR premier series championship, a first in the history of the sport.

Johnson ran three NASCAR Cup Series events during the closing weeks of the 2001 season, but when the 2002 season rolled around, Johnson was one of the sport's best prospects and top rookie contenders. To prove that, he put his Hendrick Motorsports Chevrolet on the pole for the season-opening Daytona 500, becoming just the second rookie to do so. It would take Johnson just ten races to find Victory Lane, taking the checkered flag for the first time at Auto Club Speedway in Fontana, California, on April 28, 2002, edging Kurt Busch by less than a second. Johnson would go on to score a total of three wins during his first year in the premier series, sweeping the season's races at Dover Motor Speedway en route to a fifth-place finish in the overall season standings.

During the inaugural Chase format in 2004, Johnson overcame a number of setbacks both on the track and off to finish second to Busch after a hard-fought battle over the last 10 races of the season. Despite eight wins, 20 top-fives and 23 top-tens, Johnson would finish second to Busch by just eight points, the slimmest margin in NASCAR history at that point. The close loss is even more incredible when you consider Johnson won four of the final six races of the season. It was the second straight season Johnson had finished second in the season-ending point standings, but that was all about to change.

The 2005 season would lead to another fifth-place finish in the series standings, but when the 2006 season rolled around, Johnson and his No. 48 Hendrick Motorsports team were ready to put on a show, but it would not be without controversy.

During Daytona Speedweeks in 2006, crew chief Chad Knaus was ejected from the track and suspended four races when NASCAR discovered a device that would alter the rear window to reduce rear downforce. With Knaus gone, Darian Grubb would have to serve as the substitute crew chief for the season-opening Daytona 500. The team did not miss a beat, with

Johnson taking the lead from teammate Brian Vickers with 17 laps to go to earn his first Daytona 500 victory.

"With the circumstances we've been through and the situation we're in, we overcame everybody against us," Johnson said in Victory Lane. "We were wrong in qualifying, [but] we came back through all of that and won the Daytona 500. That's something I am so proud of. We looked as bad as we could ever look, and we stepped up and got the job done."

The team would get the job done again three weeks later at Las Vegas Motor Speedway, earning their second victory of the season. Johnson would win at Talladega Superspeedway in May, and in August, he would earn his first career victory at Indianapolis Motor Speedway—his first of four wins at the iconic speedway. Throughout the regular season, Johnson led the series standings for all but four weeks but was trailing Matt Kenseth when the 10-race Chase for the NASCAR Nextel Cup got underway at New Hampshire Motor Speedway.

In uncharacteristic fashion, Johnson finished outside the top ten in the first four Chase races. However, Johnson finished second at Charlotte Motor Speedway in October to kick off a five week stretch of top-two finishes, including a win at Martinsville Speedway, his fifth victory of the season. When Johnson took command of the Chase standings after his runner-up finish at Texas Motor Speedway, he never looked back. Finishing ninth in the season finale at Homestead-Miami Speedway, Johnson earned his first NASCAR premier series championship by 56 points over Kenseth.

"I'm really looking forward to representing this sport as its champion next season," Johnson wrote in the 2006 *NASCAR Nextel Series Yearbook*. "There's a level of respect that comes with being a champion. I've given that respect to other drivers and teams, and I'm looking forward to carrying that banner all next season—being the champion and experiencing everything that comes with it."

Not only would Johnson carry the championship banner during the 2007 season, he did all he could to ensure he hoisted the title at the end of another hard-fought Chase. This time, Johnson would be forced to duke it out over the final ten races with teammate and mentor Jeff Gordon. Johnson entered the 2007 Chase with the points lead, despite losing 100 points due to a rules violation by Knaus at Sonoma Raceway.

Once the Chase got underway, Gordon proved to be a formidable foe. The veteran driver scored back-to-back Chase wins at Talladega Superspeedway and Charlotte Motor Speedway to carry a 68-point advantage over Johnson into the next race at Martinsville Speedway. That advantage would not be enough, however, as Johnson would go on to score four consecutive victories at Martinsville, Atlanta Motor Speedway, Texas Motor Speedway, and Phoenix Raceway.

Heading into the final race of the season, Johnson had an 86-point lead over his teammate. Johnson secured his second consecutive premier series championship by starting on the pole and finishing seventh at Homestead. Johnson became just the ninth driver in NASCAR history to win back-to-back titles, and the first since Gordon accomplished the feat in 1997 and 1998. But once again, Johnson wanted more.

As the 2008 season rolled around, many began to look at Johnson's back-to-back championship efforts and wonder if he could tie Cale Yarborough's mark of three consecutive championships from 1976 to 1978. Johnson and the No. 48 team stumbled out of the gate to start the season, scoring just one top-five finish in the first five races. Yet as the season continued, Johnson found his stride, winning for a second time at Indianapolis Motor Speedway. When he scored back-to-back wins at Auto Club Speedway and Richmond Raceway, Johnson entered the Chase in third place.

Once in the Chase, Johnson's biggest competition would come from Roush Fenway Racing's Carl Edwards. Both drivers earned three victories over the final ten races, but Edwards's title hopes took a hit when he was caught up in a wreck at Talladega Superspeedway and suffered an electrical problem at Charlotte Motor Speedway. Thanks in large part to a win at Phoenix Raceway in the second-to-last race of the season, Johnson once again hoisted the championship trophy at Homestead-Miami Speedway. The accomplishment made him just the second driver in NASCAR history to win three consecutive titles. Amazingly, Johnson was not done yet.

Johnson and the No. 48 team would struggle at Daytona in 2009, but the team would rebound throughout the regular season as he chased Gordon and Tony Stewart, who led the series standings for much of the season. Once the Chase kicked off, Johnson went into high gear, earning four wins, seven top-fives and nine top-tens. By taking his fourth consecutive title, Johnson set himself apart from the competition and wrote a new page in the NASCAR history books. While he became the fourth driver to win four championships—joining Richard Petty, Dale Earnhardt, and Gordon—he was the first to do so consecutively.

As the 2010 season rolled around, Johnson again started the regular season off on the right foot. Despite a 35th place finish in the season-opening Daytona 500, Johnson and his Knaus-led team won three of the first five races, taking the points lead by week six at Martinsville Speedway. However, a crash at Talladega and a tenth-place finish the following week in Richmond cost Johnson the top spot, but he remained a factor at the front through the rest of the regular season, winning a total of five times during the first twenty-six races.

Yet as the final ten Chase races kicked off at New Hampshire Motor Speedway, the championship battle would be tight between Johnson, Denny Hamlin, and Kevin Harvick. After a dominant win at Dover and second-place finish at Kansas the following week, Johnson was able to take the lead in the Chase standings. However, a win by Hamlin at Texas Motor Speedway put the Joe Gibbs Racing driver out front with two races remaining. During that race in Texas, Johnson was plagued by slow pit stops. After teammate Jeff Gordon wrecked out of the race, Chad Knaus called an audible and utilized Gordon's over-the-wall crew on the No. 48 mid-race. The change would remain in place for the rest of the Chase.

Hamlin stumbled at Phoenix, finishing 12th and losing almost half of his points lead heading into the final race of the year. Leading Johnson by only 15 points, Hamlin would have to win the race or lead the most laps and finish second if he wanted to earn his first championship.

As the final race of the season approached, the four-time champion and the edgy Harvick played mind games with Hamlin, who was distraught over his poor performance at Phoenix a week prior. The tactic seemed to work, as Hamlin qualified 37th out of 43 cars. Johnson's team was able to overcome an early issue on pit road, but led one lap and finished the season finale in the second spot behind race winner Carl Edwards. Hamlin spun through the grass early in the going and would finish 14th, while Harvick was third.

By finishing second and leading a lap, Johnson took command of the Chase standings by 39 points to earn an unprecedented fifth consecutive NASCAR premier series championship.

"We have had the highs and lows of the Chase, but to have it all come around, and to look every single one of my crew guys in the eyes on that stage tonight there's a different feeling about it," Johnson told the media after his fifth straight championship. "It is so cool. I think we were very relieved for the first [championship], and it was super, super special. But this has a different feel. And even coming in, even through the race, the final races of the Chase, I've been saying all along, I've had a good time with this. This has been fun. I was, one, so happy to be a part of three guys racing for the championship, then obviously going for five in a row. I have really soaked in this experience and enjoyed it and just so happy to come out on top."

Johnson's reign atop NASCAR's premier division would come to an end in 2011, when Tony Stewart and Carl Edwards would wage a fierce battle over the final ten races of the season, ending the year tied in points. Stewart was given the title in a tiebreaker, while Johnson finished sixth, his worst finish in the standings at that point.

The five-time champion would once again be denied in 2012, as Brad Keselowski delivered the first NASCAR premier series championship for team owner Roger Penske. Johnson would finish third.

Johnson's amazing career would inch one step closer to history in 2013, when he earned his sixth premier series championship by 19 points over Matt Kenseth. The sixth championship moved Johnson within one title of matching the high mark set by NASCAR Hall of Fame members Richard Petty and Dale Earnhardt, who both earned seven titles.

It would take another three years, but in 2016 Johnson etched his name in NASCAR history once again by securing his seventh Cup Series title, tying him with Petty and Earnhardt. Under the new winner-takes-all championship race format introduced in 2014, Johnson was in contention with Team Penske's Joey Logano and Joe Gibbs Racing's Kyle Busch and Carl Edwards.

Edwards led 47 laps, the most of any title contender, but was involved in a wreck that ended his race and championship hopes on Lap 257. The race and the championship would come down to an overtime restart with Johnson leading the inside line and Kyle Larson leading the outside lane. Logano was tucked right on Johnson's back bumper for the restart but overdrove the first corner and slid up the track and out of contention for the win. With a great restart, Johnson took the lead and led the final two laps to take the victory and the championship at the same time.

The seven-time champion retired from full-time NASCAR competition in 2020, switching gears and trying his hand at open wheel in the NTT IndyCar Series. Johnson ran a partial season in 2021 and the full season, including his first Indianapolis 500, in 2022.

For the 2023 season, Johnson turned his attention back to NASCAR. The seven-time champion joined up with fellow seven-time champion Richard Petty and team owner Maurey Gallagher, and their team Petty GMS Racing. The organization was rebranded as Legacy Motor Club for the 2023 season, with Johnson in an ownership role and part-time driver.

Johnson continues to add to his legacy in NASCAR, but one of his first major "firsts" in the sport's history was his run of five consecutive championship titles from 2006 to 2010.

Such Class: NASCAR's First Hall of Fame Class

As NASCAR moved into the twenty-first century, many of the sport's leaders envisioned the idea of an official NASCAR Hall of Fame. While multiple racing hall of fames sprinkled the country, NASCAR was the only major American sport that did not have an official hall of fame to honor its best drivers.

That all changed in 2005, when NASCAR began soliciting proposals from a number of cities interested in hosting the first-of-its-kind facility. NASCAR received initial proposals from Charlotte, North Carolina; Kansas City, Kansas; Daytona Beach, Florida; Atlanta, Georgia; Richmond, Virginia; the Birmingham/Talladega area of Alabama; as well as Michigan.

When the official bids were placed in June 2005, the choices had been narrowed to Charlotte, Kansas City, Daytona Beach, Atlanta, and Richmond.

"NASCAR is honored to receive proposals from these five cities that all play an important role in hosting NASCAR Nextel Cup events each year," said Mark Dyer, NASCAR Vice President of Licensing. "These five cities are all winners and are to be congratulated for their diligence and dedication to the Hall of Fame project. We now will get to work and study each of these proposals carefully and completely. Later this summer, we will schedule site visits to each of the five cities.

We are excited about the prospect of partnering with one of these cities to produce a world-class facility that will enshrine the legends of NASCAR and give our millions of loyal fans a touchstone of the sport they love."

While each location had its own reasons for wanting to host the official NASCAR Hall of Fame, the bid for Charlotte garnered the most support from veteran NASCAR drivers and state politicians.

After site visits and multiple proposed plans, NASCAR made its official decision on March 6, 2006, granting the winning bid to Charlotte.

"To NASCAR fans everywhere, it is my distinct honor to announce that NASCAR has selected Charlotte, North Carolina, to be the home of the NASCAR Hall of Fame," said NASCAR Chairman and CEO Brian France. "The winners in this process are the 75 million NASCAR fans nationwide, who will have a Hall of Fame to call their own. The City of Charlotte will welcome fans from around the country and even the world to the NASCAR Hall of Fame."

On January 25, 2007, NASCAR's biggest names from past and present gathered on a cold and windy morning to officially break ground on the NASCAR Hall of Fame in uptown Charlotte. Longtime motorsports radio broadcaster Winston Kelley was tapped as the Hall of Fame's executive director, while Buz McKim was named the facility's historian.

In May 2007 excavation began on the facility, which would feature the Hall of Fame and a nineteen-story NASCAR office tower. The proposed NASCAR Hall of Fame would be 150,000 square feet, with more than 40,000 square feet of interactive exhibits. The estimated cost of the facility was listed at 195 million dollars.

As construction continued, NASCAR announced on June 19, 2008, that the official grand opening ceremony would take place on May 11, 2010, with the inaugural five-member class being inducted on May 23. One month later the first twenty-five nominees were announced.

The nominees were a who's who of NASCAR history. The names included: Bobby Allison, Buck Baker, Red Byron, Richard Childress, Dale Earnhardt, Richie Evans, Tim Flock, Bill France Sr., Bill France Jr., Rick Hendrick, Ned Jarrett, Junior Johnson, Bud Moore, Raymond Parks, Benny Parsons, David Pearson, Lee Petty, Richard Petty, Fireball Roberts, Herb Thomas, Curtis Turner, Darrell Waltrip, Joe Weatherly, Glen Wood, and Cale Yarborough.

The Voting Panel would determine the five inductees after a day of debates and voting, overseen by the accounting firm Ernst & Young. The panel consisted of fifty ballots, twenty coming from the initial Nominating Committee—officials from NASCAR, the Hall of Fame, major track ownership groups, and operators of historic short tracks—and thirty from a group of former drivers, owners, and crew chiefs, as well as manufacturer representatives and media. One vote was left open for the fans, who would cast their collective votes online.

On October 14, 2009, the inaugural NASCAR Hall of Fame Class was officially announced. To the surprise of few, the five names selected were Bill France Sr., Bill France Jr., Junior Johnson, Richard Petty, and Dale Earnhardt.

"It was a very historic feeling," said H. A. "Humpy" Wheeler, former president and general manager of Charlotte Motor Speedway and member of the voting panel. "I think a lot of people remarked it was like being at the Streamline Hotel when NASCAR was founded—without all the unfiltered Camel cigarettes and probably whiskey bottles, although we could have used a few of them."

"The atmosphere was fantastic," said McKim. "We had a two-and-a-half-hour discussion time between 10 and 12:30 and we got to thinking, 'Hmm, how are we going to fill up that time?' Man, it was awesome. Everybody that was on the voting panel was so diligent about their job, they took it so seriously. Everybody had great ideas and great suggestions."

After opening ceremonies took place on May 11, 2010, an estimated 10,000 fans streamed through the new NASCAR Hall of Fame during its first week. The biggest day for the new hall would come on May 23, when the first class was officially enshrined.

In opening the ceremonies, longtime NASCAR broadcaster Mike Joy detailed the long road to the Hall of Fame and the importance of the first class.

"More than sixty years of dreaming and determined effort have brought us to this moment," Joy told the crowd. "NASCAR now enjoys its own Hall of Fame. Charlotte joins Cooperstown, Canton, Springfield, Toronto, and St. Augustine in hosting the celebrated sidelines of America's major professional sports.

"Each of today's inductees set the standard for his era on or off the track and each contributed mightily to the increased popularity of NASCAR," he continued. "Today's NASCAR Hall of Fame inaugural class will stand for all times as the Mount Rushmore of our sport."

First up was NASCAR founder Bill France Sr., the man with a vision to gather the best stock car racers of the day in Daytona Beach, Florida in 1947 to organize the sport and carry it into the future. Don Cassidy, a NASCAR insider for more than fifty years, inducted France as the first member of the NASCAR Hall of Fame.

"Bill's dreams of growth for NASCAR were only exceeded by his desire that stock car racing become a recognized and respected professional sport in America," Cassidy said. "And if he were here today, he would be the very first one to acknowledge that NASCAR has exceeded his dreams."

NASCAR's "King," Richard Petty was the next to be enshrined. The second-generation driver with 200 NASCAR premier series victories, seven championships, and seven Daytona 500s was introduced by his cousin and former crew chief Dale Inman. Richard's son, Kyle, officially inducted him into the Hall. Setting the standard for all of the drivers who followed, Petty thanked the fans during his closing comments.

"We wouldn't be here without the fans," Petty said standing on stage. "There wouldn't be a Richard Petty. There wouldn't be a NASCAR."

Bill France Jr., the second-generation NASCAR leader, guided the sport through the growing stages of the 1970s and 1980s, expanding the sport in terms of sponsorship dollars and television coverage. Team owner Rick Hendrick introduced Bill Jr.'s children, Lesa and Brian, who would induct their father as the third member of the NASCAR Hall of Fame.

"Dad was tough. There's no doubt about it," Lesa said. "He was demanding. He had every single right to be because he expected more from himself. He always did. He loved this sport. He was passionate about it. He built it from the ground up. When I say 'the ground up,' I'm talking about a backhoe at Daytona International Speedway."

Considered by many to be one of NASCAR's greatest drivers behind the wheel, Junior Johnson was a former moonshiner that rose to the greatest heights of NASCAR success, becoming the fourth member of the inaugural Hall of Fame class to be inducted. When it came time for contractors to install the moonshine exhibit in the new Hall, Winston Kelley contacted Junior Johnson to ensure its accuracy. Instead of walking them through the process, Johnson showed up with a pipe wrench and a pair of Channellocks to help assemble it properly.

Former champion and eventual Hall of Fame member Darrell Waltrip introduced his old team owner by sharing numerous stories about Johnson's success, innovations and style of leadership.

"Junior was an innovator. He always thought outside of the box," said Waltrip. "The things that he did were first. Let me tell you, if you want to make Junior Johnson happy, just do something first before anybody else does. Junior was not a follower. Junior was a leader."

Finally, team owner Richard Childress took the stage to introduce and induct his longtime friend and former driver Dale Earnhardt as the fifth member of the inaugural NASCAR Hall of Fame Class.

"Dale carved out his own piece of NASCAR history," Childress said of his late friend. "He took the sport to another level. In that process, he brought millions of fans along for the ride."

After Earnhardt was officially inducted into the Hall of Fame, his family took the stage to remember the man who dominated NASCAR in the 1980s and 1990s before losing his life on the final lap of the 2001 Daytona 500.

"Dale Earnhardt was a man who personified the American dream," said Teresa Earnhardt, his wife. "He worked hard. He earned everything he had and he enjoyed it. This is an achievement of a lifetime. To be able to celebrate it, for me this is a moment of pride for Dale that I just can't put into words."

Dale's children, Kerry, Kelley, Dale Jr., and Taylor all spoke of their father's illustrious career and personal life, but Teresa ended the ceremony with a final quote from the man they called "The Intimidator."

"A racer wants to race and win," she said. "Imagine having the opportunity to do that for a living, and then to be successful, and then to be considered one of the greatest drivers that ever raced, especially by a group of peers. It's one of the greatest honors a driver could ever receive. I've had a great career. If it ended tomorrow, I'd have no regrets. Dale Earnhardt."

The ceremony to induct the first class of the NASCAR Hall of Fame was a monumental day for the history of the sport. A dream to organize the ragtag stock car racing groups in 1947 had paid off for Bill France Sr., and through his vision and the efforts of each inductee, the sport grew and flourished into one of the world's top spectator sports.

A New Form of Ownership— The Charter System

For much of NASCAR history, if someone wanted to own a race team and go racing in the sport's premier series all they had to do was raise the capital, find a car and driver, obtain a NASCAR license, and go racing. There were no safety nets, no guarantees, and a lot of risk.

NASCAR served as the sanctioning body, while teams and drivers were seen as independent contractors. Under this format, NASCAR owed team owners and drivers little when it came to financial security or stability looking to the future.

Despite multiple efforts by drivers throughout the sport's history to organize and unionize, NASCAR remained in charge with very little indication things would change one way or another.

Yet in 2016, things changed dramatically for team owners in the NASCAR Cup Series.

On February 9, 2016, NASCAR Chairman and CEO Brian France, other NASCAR leadership, and some of the sport's top owners announced the creation of the Charter system. By working closely with the Race Team Alliance—an organized group of NASCAR team owners—the system was put in place to provide some assurances to the owners and a bit of a safety net to their investment in their teams and the sport. Negotiations between NASCAR and the RTA lasted eighteen months before reaching an agreement.

NASCAR issued 36 total Charters to teams. Each Charter would be used for a single car, so if an organization is a four-car team such as Hendrick Motorsports they would have four total Charters. The Charters guarantee that team entry into every NASCAR Cup Series points-paying event, allowing them a guaranteed portion of the race purse each week. Each organization was also capped at four cars and four Charters per team.

Teams can sell the Charter, and they are also able to lease it to another team for one full season. If an organization leases their Charter to another team, they can do so only once in the first five years they own the Charter.

While there are guarantees for teams that own a Charter, there are also performance metrics in place to ensure they can keep their Charter moving forward. If an organization finishes in the bottom three of the 36-Chartered teams for three straight years in the season-ending owner points, NASCAR has the ability to revoke the Charter.

"Where we ended up is a very good place because we believe we made the team owner model more reliable, more stable, more open, more open to new investors, more capital to come into the sport," France said. "We've done that in a way that preserves a lot of the traditions that are important to NASCAR: the competition level, et cetera, how we qualify for events, the fact that if you have the talent and the aspiration to join NASCAR, you can do that in the new model as you could in the old.

"In the end we have a model that fits the 21st century and I think will serve us well into the future. If it works properly, and we're very certain that it will, it's going to do some things," he continued. "It's going to create more opportunities for, as I said, new owners to come into the sport. That's number one. Number two, it's going to have the ability to field better, more competitive race teams. That's what our fans want. They want to see their favorite driver have an opportunity that's robust and competitive and all the things that makes a great race team succeed."

Rob Kauffman, at the time co-owner of Chip Ganassi Racing, was instrumental in working with NASCAR to develop the system and work out all the details through lengthy negotiations and meetings with all parties that lasted nearly eighteen months. The group not only looked at the traditional model that was in place for nearly six decades within NASCAR, but they also looked to other sports such as the NFL, Formula One, cricket, etc., to take all they could learn and apply it to the new model being ironed out.

Proud of their work to get to announcement day, Kauffman made it clear this was just the beginning of something that would require more work and cooperation between NASCAR and team owners moving forward.

"I think we focused on three key principles: participation, economics, governance. That's the foundation stones for how this new system works," said Kauffman.

"I think what we've done here is now we've put the teams on a more stable footing," he continued. "What does that mean? I think it really allows the teams to invest for the future. If you're living year to year, you can't really think about the long-term. Here what you're able to do is say, Okay, I'm going to take this risk on this young driver, see how they work out, develop this crew chief, this crew member, even some of the technical sides, even your basic business infrastructure. It's a step in the right direction. I think all the work starts here."

Part of the sports since the earliest days, Richard Petty was on hand for the announcement of the Charter system, calling it "the second most important thing that's ever happened in

NASCAR" because it allowed NASCAR and the team owners to come together and work to find solutions to make the sport successful long into the future.

"We don't understand how NASCAR works, okay? [NASCAR] don't know how the race teams work. So now we're going to be in the same room talking about the same problems and solving the problems together," said Petty. "From that standpoint, it's going to be one of the greatest things that's ever happened to NASCAR and Cup racing. Like I say, I've been here a long time, I've seen a lot of stuff go under the bridge more or less. But this is the next generation of what we're going to have to do to make racing that much bigger."

The implementation of the Charter system was welcomed by nearly every aspect of the sport—something that is quite rare in NASCAR history. With teams primarily dependent on sponsorship dollars to keep their organizations afloat year-to-year, the security of the Charter allows them a buffer to work within and security to know that if a team owner wants to or needs to exit the sport, they may sell their Charter on the open market to the highest bidder. The system was set to expire in 2024, with hopes of renegotiating and extending it into the future.

Although introduced in 2016, the Charter system still did not prevent the 2017 NASCAR Cup Series championship team, Furniture Row Racing owned by Barney Visser, from leaving the sport the following year.

Based outside the heart of NASCAR country, Visser's team was in Denver, Colorado, and operated a bit on an island compared to other NASCAR Cup Series organizations. Despite owning one of the 36 Charters issued by NASCAR, Visser cited rising costs and little return on investment as his main reasons for leaving the sport right after winning the title.

"This is not good for anybody," Visser said in a statement. "The numbers just don't add up. I would have to borrow money to continue as a competitive team and I'm not going to do that. This was obviously a painful decision to arrive at knowing how it will affect a number of quality and talented people."

Visser initially sold the Charter of the No. 77 car to JTG Daugherty Racing, which allowed them to expand to a two-car operation in 2018. After announcing the team would cease operations after the 2018 season, Visser sold his final Charter to Spire Motorsports for their No. 77 Chevrolet.

As the Charter system took hold and the sport adjusted, team owners still came and went but they were able to capitalize on the rising cost of the Charters to cut their losses from closing the team. With only 36 Charters issued, finding an available Charter to purchase or lease became a hot commodity.

By the 2020 season, Charters were being sold for reportedly $6 million. One year later, Spire Motorsports sold two Charters to Kaulig Racing, fetching around $10 million for one and around $8 million for the other. In 2022, the *Sports Business Journal* reported the price of available Charters had climbed as high as $20 million.

With the value of Charters rising, a team owner can now walk away from the sport without losing everything, which has been the case for most of NASCAR history. If they choose to walk away, a guaranteed starting spot in the NASCAR Cup Series and portion of the purse each race is a big incentive for interested parties not just within motorsports but all walks of life.

One aspect of the Charter system that worked according to plan is bringing new owners to the sport. Since its introduction, the NASCAR Cup Series has seen an influx of new ownership in the form of basketball legend Michael Jordan and international music icon Pitbull, both of whom were involved in starting new Cup teams in ahead of the 2022 season.

Jordan partnered with three-time Daytona 500 champion Denny Hamlin to form 23XI Racing, with Bubba Wallace as their driver. Pitbull, whose real name is Armando Christian Perez, partnered with racer Justin Marks to purchase the assets and Charters from Chip Ganassi Racing to form Trackhouse Racing with Daniel Suarez, the first Mexican-born NASCAR Xfinity Series champion, as their driver. Both teams eventually added a second Chartered team, and both have found Victory Lane multiple times.

"This acquisition provides Trackhouse a platform for years to come, enabling us to field multiple teams, elevate our on-track performance, deliver great value for our sponsors and partners, build lasting community impact programs and continue to grow our brand beyond the sports conversation," Marks said.

While the Charter system brought new ownership and stability to existing owners, its original expiration date was 2024. Throughout the 2022 and 2023 season, the RTA and NASCAR were once again at the negotiation table with Dave Alpern of Joe Gibbs Racing, Curtis Polk of 23XI Racing, Steve Newmark of RFK Racing, and Jeff Gordon of Hendrick Motorsports leading the negotiations for the RTA.

This time, with a new TV package looming, the owners wanted to carve out a larger portion of the money brought through the broadcast of the race and extend the Charter system to make it a permanent fixture in the sport.

In his annual State of the Sport Address at the end of the 2022 season, NASCAR President Steve Phelps described the Charter system as being "very good for NASCAR," pointing out that the guarantees of a Charter certainly help attract new ownership interest.

"Do I think we'll extend the charters? I do. Do I think it's a good thing for the sport? I do," he said.

When asked about letting the Charter agreement expire, Phelps admitted, "There's certainly that possibility."

On April 5, 2023, NASCAR was set to meet with team owners for a regularly scheduled quarterly meeting. Instead, none of the team owners showed up to the meeting. Although they claimed it was not an outright boycott, they clearly sent a message to NASCAR.

Hamlin, co-owner of 23XI Racing, indicated that the negotiations had deteriorated and become worse than where they had been just a few months prior. "As team owners, we feel like time is running out," he stated. "The Charter agreement is not directly linked to the TV deal. So, they could be done at different times. . . . The further NASCAR pushes us back to our timeline of our Charters [expiring], the less time we have to get a deal done. If we don't get a deal done, then we could be racing all next year with lame duck Charters for a year. We'd be obligated to race and lose them the following year."

Most of the frustration from the RTA negotiating party at the time was due to the lack of involvement of NASCAR Chairman and CEO Jim France and NASCAR Executive Vice Chair Lesa France Kennedy.

As of this publication, the negotiations are still ongoing with no real end in sight as of yet. It is clear the Charter system has worked in many regards, but whether it is something that remains into the future of the sport, that is history yet to be written.

A Season Without Fans

ONE THING that NASCAR has always prided itself on is how accessible the sport is to fans. Unlike most sports, fans attending a NASCAR race have more access to the stars of the sport right up until the time they climb in the cars, fire the motors, and start the race. Packed grandstands, infields full of campers and tents, throngs of people flooding through the fan zones and parking lots, these are the norms week-in and week-out at NASCAR events across the country.

That was certainly the case when the 2020 NASCAR Cup Series kicked off the season with a packed house for the 62nd running of the Daytona 500. Sitting US President Donald Trump was on hand for the event, the grandstands and infield were packed to the brim with a sell-out crowd—and were they in for a show.

The race featured several dramatic moments and thrilling racing throughout, but a spectacular crash coming to the checkered flag saw Ryan Newman's car fly through the air and be hit in the driver side window by Corey LaJoie. It a silence over the massive crowd. While Denny Hamlin tamely celebrated his third Daytona 500 victory, safety workers carefully cut Newman from his car and transported him to a local hospital.

While many feared the worst, Newman defied the odds and walked out of the Halifax Medical Center in Daytona Beach, Florida, less than forty hours after his crash. The 2020 season started off with one of the most dramatic and uncertain moments in recent NASCAR history, but few could have predicted just how uncertain and trying the year would be.

After four races, Team Penske's Joey Logano already had two wins, Newman was on the road to recovery, and the season was shaping up to be one of the most competitive and exciting in years. As the series headed to Atlanta Motor Speedway for the fifth race of the season, things came to an abrupt and unprecedented halt.

For those within NASCAR, the sport is life and there is very little separation. The NASCAR Cup Series schedule spans from February through November, there are few days off, and the work does not stop during the off-season months. In essence, racing is life.

In March 2020, life as everyone across the world knew it changed when the coronavirus (COVID-19) ran rampant and turned into a global pandemic. With the virus spreading at an alarming rate and death rates rising, countries across the globe took extreme measures by locking down in an attempt to stop the spread of the virus.

The NASCAR world was en route to Atlanta Motor Speedway when this took place. Haulers carrying the cars had reached the speedway and the planes carrying teams to the track were in the air to the area. However, on March 13, NASCAR announced it was postponing that weekend's events, as well as the following week's scheduled race at Homestead-Miami Speedway.

The teams turned around and returned to North Carolina with a host of questions and a lot of uncertainty about what would come next, most of which was beyond their control. Not only would teams not be allowed at the racetrack, working at the race shop and preparing race cars was now off the table for the time being as well, as lockdown procedures were put into place.

"We're navigating this process with the entire industry and look forward to providing further details when they are finalized," NASCAR president Steve Phelps said days later. "We're working through both the complexity of our sport and our many industry stakeholders as well as the complexity of this pandemic and its impact on our daily lives.

"These clearly are unprecedented times with information changing by the hour. Collectively our industry has made several difficult decisions, all with one thought in mind: the health and safety of our fans, our competitors, employees, and everyone in the industry," he said. "The situation we are facing transcends the world of sports. What is most important now is we take precautions to keep everyone as safe as possible during these challenging times."

Phelps was clear NASCAR would keep safety at the forefront of everything, but the goal was to run a complete 36-race season and crown a champion for the 2020 season. How they would accomplish that would require flexibility, cooperation, and understanding from all parties involved.

As the world adjusted to the new reality of lockdowns, NASCAR shifted to iRacing to fill its need for action. The sport organized virtual races among the NASCAR Cup Series drivers and teams using the PC-based simulation program. Partnering with FOX Sports, NASCAR ran seven virtual races and broadcast them to the public, with interviews and interaction from the drivers—who were racing from the safety of their homes.

While the virtual races were entertaining, everyone was itching to get back to the racetrack as soon as possible.

On April 30, 2020, NASCAR announced it would return to in-person racing starting on May 17. Working closely with federal, state, and local officials, the sanctioning body laid out a four-race schedule for the NASCAR Cup Series, as well as two

Xfinity Series and one Camping World (now Craftsman) Truck Series races. The events would be held at Darlington Raceway and Charlotte Motor Speedway, and no fans were permitted at the events.

Only a limited number of personnel were allowed at the track, each person was required to fill out a health questionnaire and temperature screening, and everyone had to wear a cloth mask to be present. Social distancing was implemented, and strict rules were in place for anyone on-site.

"We have a lot of confidence in our plan. We know we have to work together as an industry to keep our own folks safe, to keep each community safe," said NASCAR vice president of racing operations John Bobo. "But it is the discipline and the safety culture of NASCAR. We're the organization that puts cars on the track four days a week at 200 miles an hour. We think it's that same discipline and eye towards safety that everybody in our industry has that is going to help us execute on this."

The events at Darlington did not feature any practice or qualifying, as the starting lineup and pit selections were determined by random draw and Charter team owner points.

On May 17, 2020, NASCAR was officially the first sport in the United States to resume after the outbreak of COVID-19. The Real Heroes 400 at Darlington Raceway was kicked off when a group of healthcare workers virtually gave the command to fire engines. Each car in the field also carried the name of a healthcare worker on the car's name rail.

The field took the green flag and for a few hours the world was back to some sense of normalcy as forty NASCAR Cup Series cars raced around the sport's oldest speedway.

In the end, Kevin Harvick and his Rodney Childers-led Stewart-Haas Racing team dominated the event leading 159 of the 293 laps to score the 50th NASCAR Cup Series victory of his career. Climbing from the car on the frontstretch after a massive burnout, Harvick was met with near silence and a completely empty grandstand.

"The weirdest part of the day for me was getting out of the car and not hearing anybody cheering. When the engines are running, you're kind of in your own little world in the car," said Harvick. "Look, I've been around this deal for a long time. This is not like anything I've ever experienced. I can tell you it's very similar to coming back after 9/11. That day had a hundred thousand fans in the stands, now you have no fans.

"The practice and in-the-car part are something we can figure out. The not having any fans in the infield, for me that took some of the enthusiasm of the win away just because of the fact you weren't able to celebrate with the crowd," he added. "For me, that was really awkward because I feed off of that stuff, enjoy those types of moments. For me. I didn't really know what to say after the race because it's such a unique situation that we're in."

Following the race, O'Donnell met with the media virtually and thanked the local healthcare workers and frontline workers for their efforts in returning NASCAR to action. He also pointed out that NASCAR would work as close as possible with other sports leagues to share practices and knowledge from what they learned returning to action amid the COVID-19 pandemic.

"We're a little bit unique in how we're able to go racing in terms of race cars, helmets, a lot of that equipment that's already in place. So, a little bit of an advantage in terms of our athletes as well, all getting them to one race," he said. "I think everybody was really looking forward to seeing how this works. Anything we can share with other leagues, we want to do that. We certainly wanted to get out and try and showcase our sport early."

The sport continued to adjust and adapt to the ever-changing situation surrounding COVID-19, the different restrictions and rules put in place in different states, running midweek races, and doubling up at tracks in states that allowed the sport to race.

On June 4, 2020, NASCAR announced the schedule that would get them through the beginning of August, while uncertainty still remained about states opening up and how best to move forward.

Days later, on June 9, they announced a limited number of race fans would be allowed back at the events starting with the June 14 race at Homestead-Miami Speedway and the June 21 event at Talladega Superspeedway. A total of one thousand fans would be admitted to Homestead, while Talladega would open to five thousand fans. All fans attending the events would be required to wear masks, socially distance, take a health screening prior to entering the facility, and more precautionary procedures.

"We have tremendous respect and appreciation for the responsibility that comes with integrating guests back into our events," said Daryl Wolfe, Executive Vice President, Chief Operations and Sales Officer, NASCAR. "We believe implementing this methodical process is an important step forward for the sport and the future of live sporting events. The passion and unwavering support of our industry and fans is the reason we race each weekend, and we look forward to slowly and responsibly welcoming them back at select events."

By August 6, NASCAR announced the remainder of the 2020 schedule for all three series, with NASCAR Championship Weekend taking place November 6-8. Safety protocols remained in place for all events and the presence of fans was determined on a race-by-race basis, according to state and local regulations.

Several drivers, including Jimmie Johnson and Austin Dillon, tested positive for COVID-19 throughout the year and were

forced to sit out of competition until a negative test was produced.

Despite all the unprecedented challenges it faced, NASCAR
accomplished its ultimate goal of completing all thirty-six races
and doing so by November 8. Limitations remained in place
on the number of fans allowed to attend each event, but in the
end, there were fans in the stands when Chase Elliott scored
the victory in Phoenix to earn his first NASCAR Cup Series
championship.

"It's just unprecedented in the history of our country, in the
history of sports, and certainly in the history of our sport. I
would suggest this is the single most difficult year that we've
faced as a sport," Phelps said when addressing the media during
championship weekend. "But through it all this industry, I've
said this before, I believe this industry does adversity better than
any sport. If you think about it, we're at a competitive disadvan-
tage. We don't own ourselves. We're not franchised, right? We
have independent contractors who come to race as one. What
we have done during this global pandemic is I think nothing
short of remarkable."

The 2021 season began with COVID-19 restrictions in place
and limited access for fans at the Daytona 500. NASCAR proved
it was not taking the restrictions lightly when they fined team
owner Chip Ganassi $30,000 and barring him from the follow-
ing week's race at Homestead-Miami Speedway for violating
the protocols in Daytona.

Eventually during the 2021 season, things began to return
to a sense of pre-pandemic normalcy as COVID-19 restrictions
such as masks and social distancing were reduced or eliminated
completely as requirements for entry.

Despite facing many unprecedented obstacles and challenges,
NASCAR came together as an industry like never before in

its history to accomplish something truly remarkable by being the first major sport back to action after the outbreak of the COVID-19 pandemic.

"One of the things that I challenged the industry to do is to continue to move as one company, as one entity, with one voice," Phelps said ahead of the 2021 season. "The industry's doing that. Racetracks working with the sanctioning body that are not our own tracks, the relationship we have with our teams, our OEMs, our broadcast partners, it's never been better. The trust that exists today versus where we were two years ago, three years ago, we are in a very, very different place."

NASCAR Goes Next Gen

FROM THE earliest days of NASCAR, drivers and teams built and prepared their own cars to show up and race each weekend. With ingenuity, creativity, and a blurred interpretation of the rule book, cars were uniquely built and set up with NASCAR only getting involved in the at-track and post-race inspection process. The series would set the rules and parameters for the cars, but it was up to the teams themselves to design and build the cars.

For the 2022 season, that drastically changed for the first time in NASCAR history when the Next Gen, or Generation 7, Cup Series chassis was introduced. To control the rising costs associated with building race cars, NASCAR developed a purpose-built chassis that each team would be required to use. All parts came from a single supplier and each team was required to use that supplier.

There were improvements aimed at overall driver safety, as well as radical changes in the transmission and gearing of the car, and the steering and braking components of the car.

The project began in 2019, with NASCAR and its team of engineers working closely with the OEMs (Chevrolet, Ford, and Toyota) to create something unique from its predecessor, sleek and safe in design, and more like modern passenger cars of the day in both looks and components. Originally planned for an introduction in the 2021 Daytona 500, the COVID-19 pandemic delayed the new car's implementation by a full year.

Over a two-year period, the car was developed and designed by a team made up of NASCAR and Richard Childress Racing engineers. The new car first hit the track at Richmond Raceway for a two-day test. On October 9, 2019, Austin Dillon was the first to test NASCAR's Next Gen car.

"This is an important milestone for the Next Gen car and the future of stock car racing," said John Probst, NASCAR Senior Vice President of Innovation and Racing Development. "There are so many new systems on the car from the front to the back that our main goal with this test was to log laps and put miles on them. The test has met—and even exceeded—our expectations, and we are well on our way to developing the final iteration of the car."

The car was also tested at Auto Club Speedway, Charlotte Motor Speedway (both the 1.5-mile oval and 2.28-mile Roval), Darlington Raceway, Daytona International Speedway, Dover Motor Speedway, Homestead-Miami Speedway, Martinsville Speedway, Phoenix Raceway, and even Bowman Gray Stadium— where Clint Bowyer, Dale Earnhardt Jr., and Tony Stewart drove the car with Larry McReynolds helping make adjustments.

Following the test, Earnhardt was fully behind NASCAR's move to the new car.

"I think this car is a great match for this track or any track, whereas the car we have now doesn't work everywhere and is difficult in places like this," he said. "I'm sold now. I was skeptical, but I think it's going to work really well."

There was plenty to adjust to, as the Next Gen car was a true departure from anything seen before in NASCAR.

The tires on the Gen 7 car would be 18 inches, wider than what was previously used in the NASCAR Cup Series, with forged aluminum wheels secured to the car with a single lug nut, as opposed to the traditional five-lug wheels used prior.

In addition, rack and pinion steering was introduced to replace the traditional steering boxes, independent rear suspension did away with the track bar, hood louvers were added to increase engine performance, a rear diffuser was introduced under the rear of each car to help with "dirty air" while in traffic, and the underbody was sealed to address the same issue.

Increased connectivity and technology in the car also meant each driver now had multiple in-car cameras looking out the windshield and back at the driver in his seat, as well as a digital camera rear-view mirror. The traditional mirror remained, but an additional digital version provided drivers with another option when racing against the competition.

One thing learned quickly during testing was the interior of the cars were very hot. To help alleviate this issue, additional ducts were added to the car, primarily at the base of the roof and front windshield, to provide more air to the driver and help cool the cockpit.

The Gen 7 car also saw the introduction of a transaxle, which combines the transmission and rear gears into one package located in the rear of the car. Gone was the H-pattern shifter, transmission, and rear-end gear, and solid-axel rear suspension used since the sport's earliest days. Now, drivers would shift by pushing the shifter forward and backward to upshift and downshift, and a fifth gear was added for the first time in the sport's history.

One aspect of the Gen 7 car was that all parts would come from 26 different suppliers. The goal of this move was to make it easier for everyone to source the parts needed to build the car and reduce the costs associated with building the parts themselves.

However, as the 2022 season approached and the debut of the Gen 7 car was nearing, parts and pieces were hard to come by for nearly every team in the sport. In many cases, supply chain issues put a strain on teams and by the start of the season few teams had the number of chassis they originally planned to have by the time.

Speaking with members of the media on January 11, 2022, team owner Richard Childress said his team had two cars for each of his two teams less than a month before the car's debut.

"You always got some concerns, but I think NASCAR is approaching it right from a safety standpoint. After each crash test they would go back and make changes to make the car safer and better. I think that's why we got kind of behind was because NASCAR's number one thing was safety with this car," he said.

The teams also faced two huge unknowns to start the season with the new car—the first ever Clash at the LA Coliseum and the season-opening Daytona 500. Both events offered opportunities to crash cars and strain the already limited amount of equipment.

The concern was so great there were discussions about not keeping the winning Daytona 500 car in the Motorsports Hall of Fame of America at Daytona International Speedway, a tradition started in 1997. In the end, Roger Penske agreed to give the No. 2 Team Penske Ford to the museum after Austin Cindric became the first rookie to ever win the Daytona 500.

While teams were concerned to start the season, once things got rolling the supply chain issues eased and inventory began to stock up. Concerns about the Next Gen car did not end there, however, as several safety issues raised red flags for everyone in the garage.

The design of the Gen 7 car, and in particular the Ford Mustang, saw an area just behind the front tires in the wheel well in which rubber often collected throughout a race. The exhaust pipes from the motor ran just above this area, which was accumulating more and more rubber each lap. On several occasions, the rubber build-up grew so hot next to the exhaust it caught fire during the race.

During the opening race of the NASCAR Playoffs at Darlington Raceway, Kevin Harvick was forced to retire his car when the rubber in the wheel well of his No. 4 Stewart-Haas Racing Ford caught fire and ended his day. The car had no other mechanical issues and Harvick was eliminated from playoff contention after the first round.

"The car started burning and as it burned, the flames started coming through the dash. I ran a couple laps and then as the flames got bigger it started burning stuff up and I think right there you see all the brake fluid that was probably coming out the brakes and part of the brake line, but the fire was coming through the dash," a frustrated Harvick said. "What a disaster for no reason. We didn't touch the wall. We didn't touch a car and here we are in the pits with a burned up car and we can't finish the race during the playoffs."

The new composite bodies on the Gen 7 made the cars much tougher and less susceptible to taking damage, which meant they could take a beating and keep on going. The drivers, on the other hand, were feeling the hits more significantly than the previous generation car, especially in a collision from the rear of the car.

With the transaxle sitting at the rear of the car behind the firewall, the rear clip was reinforced to be strong, rigid, and able to take a large impact. Instead of the rear of the car taking the brunt of the impact's force, the drivers were increasingly learning they took most of the force.

On July 23, 2022, during a qualifying session at Pocono Raceway, the rear of Kurt Busch's No. 45 23XI Racing Toyota came around exiting Turn 2 and made contact with the outside wall with the right rear of the car before it snapped around and hit with the right front. Busch suffered a concussion and Ty Gibbs served as his replacement driver for the remainder of the 2022 season.

The 2004 NASCAR Cup Series champion with thirty-four career wins, Busch was forced to walk away from full-time competition due to his symptoms from the concussion he suffered earlier in the year. He made the difficult announcement on October 15, 2022, at his home track of Las Vegas Motor Speedway.

Busch was already locked into the NASCAR Playoffs thanks to a win at Kansas Speedway in June. Although he worked to get back in the car, the effects of his injuries were too much to overcome to remain competitive behind the wheel.

"With head movements, and torso movements with my heart rate elevated, that's when things move quick in my peripheral," Busch said, a little under three months removed from his accident at Pocono. "It would be like looking in the mirror, and then back towards the windshield and around at the competitors, checking the dash like just things are moving quick. And things are slowing down. Things are coming back to me. I just know I'm not 100 percent. The vestibular side is really where I'm focused on with my concussion."

During the September 25 race at Texas Motor Speedway, two drivers were involved in separate incidents that led to injuries and forced them out of the car. Alex Bowman suffered a concussion when his No. 48 Hendrick Motorsports Chevrolet lost the right rear tire, slid up the track in Turn 4, and hit the rear of the car, then slapped the wall with the right front. While the damage to the car did not look too severe, Bowman certainly felt the hit inside the car.

"Done. I can't drive the rest of the day," Bowman told the crew on the radio after the hit. "I don't understand how this thing is still rolling. That's the hardest I've ever crashed anything in my life."

Bowman brought the car to pit road and the team made repairs. He finished the race in 29th, the last car running.

During that same event, Cody Ware lost control of his No. 51 Rick Ware Racing Ford in Turn 4, hitting the outside wall hard head-on, then speeding down and hitting the pit wall. Ware climbed from his car and fell to the ground twice as he was attended to by safety and medical personnel.

Ware was cleared to race the following weekend, despite suffering a fractured ankle and ligament damage in the incident. When he showed up at Talladega Superspeedway the next week, Ware was sporting a crutched and swollen ankle but still raced.

Bowman, on the other hand, was forced to sit out of competition five out of the final six races, ending his chances at a championship. Within months, two playoff contenders were sidelined with injuries sustained in rear-facing hits with the Gen 7 car.

One week following these incidents, Bowman's teammate Chase Elliott spoke out about his frustrations with the Gen 7 car and the injuries that largely resulted from the car's design.

"I think it's just super surprising to me that we allowed that to happen, but we did, and now it's just about how do we go forward from here making sure we make the right choices to improve what we have and keep things like what happened to Alex this week from happening, and what happened to Kurt," the usually reserved Elliott said prior to the race at Talladega Superspeedway.

"We had [plenty of] time to test this car and crash it and do all the things we need to do to make sure some of these things aren't happening that are happening now," he said. "We had a ton of time to do that, and this car was delayed an entire calendar year on top of that. You have to think, we had an extra year of time to work on it, and we're still in this position. There's no excuse for going backward. We have too many smart people, too much technology, too many years of crashing and racing at all these same racetracks to have some of these things that are going on that are. Test next week or no test next week, we should not be in the position that we are in. When you come out with a new product, you should take steps forward, not stay the same or go backward."

Hamlin called on NASCAR to abandon the project and start over with a completely new design.

"The car needs to be redesigned; it needs a full redesign," Hamlin said. "It can still be called Next Gen, but it needs to be redesigned everywhere—front, middle, rear, competition. The whole thing needs to be redesigned. We've got a Martinsville race coming up and it's going to be tough. This thing is going to be exposed about how bad the race is and that's just part of it.

"Competition, safety, we'd like to have it all better. But certainly, we just took a step back in safety and competition this year," he said.

Hamlin confirmed concerns from many drivers about the car's safety after learning the results of multiple crash tests prior to the car's introduction.

"We threw up red flags over a year ago and they just didn't respond. They kept pushing that this car needs to be on track at all costs," he said.

Prior to the race at Talladega, Steve Phelps met with Hamlin privately to discuss his concerns about the car and comments in the media. On October 9, 2022, NASCAR met with drivers as a group to discuss ways they were working to improve the safety of the car and shared data from a crash test with updated components planned for the 2023 season.

Coming out of the hour-plus meeting, most of the drivers described the atmosphere as tense and a situation where drivers aired their frustrations with NASCAR's leadership. Phelps admitted the sanctioning body should have called an all-driver meeting earlier than it did but called the initial meeting "terrific."

"Safety is the single-most important thing for NASCAR, the safety of our drivers," Phelps said in an interview with Marty Snyder, Dale Earnhardt Jr., and Jeff Burton on that weekend's pre-race broadcast on NBC Sports. "We have a two-decade history that would suggest that that is a true statement. Are there things that we need to do to this race car to make it safer? Yes, particularly in the rear of the car. But there are also things in this car that are safer than the last car. So, we're going to continue to iterate on the car working with our drivers."

NASCAR continued to meet with the drivers as a group and through a Driver Advisory Committee to address the safety of the Gen 7 car and ways to improve the product to reduce the force absorbed by the driver in crashes.

"I think the communication between the sanctioning body and the drivers over this past five or six weeks has completely shifted the narrative on how the drivers are feeling about the area of safety or raceability, whatever it is the concerns are, the conversations we're having with the drivers, you can tell there's a difference, right, in how the drivers are speaking even to all of you," Phelps told the media in his State of the Sport address on November 4, 2022.

Despite the safety concerns, one area in which the Next Gen car lived up to its hype was creating more parity.

Chase Elliott led the series with five wins, while season champion Joey Logano earned four victories en route to his second NASCAR Cup Series championship. However, a total of nineteen drivers went to Victory Lane throughout the 2022 season, tying a record for the most drivers to earn a win in a season: joining 1956, 1958, 1961, and 2001.

According to NASCAR Loop Data, the Gen 7 car broke records for green flag passes for the lead (1,544) and the highest percentage of lead lap finishers in the modern era (59.46 percent), but also saw an overall year-over-year increase in green flag passes, up nearly 6.4 percent.

"The introduction of the Next Gen car that was so important. If you consider that before this year, the Next Gen car, you had to have a relationship with one of five race teams if you wanted to come into this sport. You had to," Phelps said. "This car changed that. What does this car do? There was a relevance to this car for OE partners. The styling was fantastic of this car. Then the question would be, Well, what's the raceability of the car? The raceability of the car was such that it resulted in nineteen different winners, so more than half the field won a race in NASCAR this year. Five first-time winners. More passes throughout the field in a single season."

The car saw more adjustments aimed and increased safety and better on-track performance for the 2023 season. The rear clip was redesigned to reduce the number of right angles and provide more areas of crushability while in a rear-facing incident.

"We took the rear bumper struts," said Dr. John Patalak, NASCAR vice president of safety and engineering. "They're made out of aluminum, and so they're made out of thinner gauge than they previously were. So, they collapse at a lower load; their buckling loads are lower.

"And then the rear clip, we removed and/or replaced structures with different cross sections, meaning we weakened the structures so that they would deform under less load. And you have to be very specific and careful where you do that so that we don't introduce problems in other circumstances where you don't want it to bend or where you can't have it bend," he said.

Throughout the 2023 season, the Gen 7 car continued to raise safety concerns as drivers were once again forced to sit out with concussion-related symptoms because of impacts. NASCAR and its team of engineers worked to address the concerns of drivers and team members, yet critics and skeptics remained in the garage. While the purpose-built Gen 7 car may have its flaws, the sanctioning body remains committed to improving both the performance and the safety of the car as it heads into the future.

A New Generation NASCAR: Clashing at the LA Coliseum

BEN KENNEDY grew up in the shadows of giants. A family legacy unlike any other in the sport, Kennedy's mother (Lesa France Kennedy), uncle (Brian France), grandfather (Bill France Jr.), great uncle (Jim France), and great-grandfather (Bill France Sr.) shaped NASCAR and built it to the successful industry it is today.

As the fourth generation of the France family grew up, Kennedy took a very familiar route. On August 28, 2010, at Gresham Motorsports Park in Jefferson, Georgia, Kennedy made his first NASCAR K&N Pro Series East start. A ruptured oil line after 55 laps ended his day after starting 12th in the No. 96 Chevrolet owned by Mark Martin.

Kennedy earned his first-career NASCAR win in dominant fashion on April 13, 2013, at Five Flags Speedway in Pensacola, Florida. Starting from the pole, Kennedy led all 150 laps to score the victory. He went to Victory Lane after leading the final 27 laps later that year at the iconic Bowman Gray Stadium in Winston-Salem, North Carolina.

After a stint of five races in 2013, Kennedy went full-time, racing in the NASCAR Craftsman Truck Series in 2014.

On August 17, 2016, the fourth generation France earned his first career win in NASCAR's top three touring series, taking the checkered flag at Bristol Motor Speedway. Kennedy edged out Brett Moffitt by 0.274 seconds. The field included NASCAR Truck and Xfinity Series champions such as Moffitt, Christopher Bell, William Byron, Matt Crafton, Daniel Hemic, Tyler Reddick, Ben Rhodes, and Johnny Sauter.

Kennedy made seventeen NASCAR Xfinity Series starts between 2016 and 2017, earning a best-career finish of 4th at Talladega Superspeedway on May 6, 2017, driving for Richard Childress Racing.

In addition to his career behind the wheel, Kennedy also owned a race team that competed in the ARCA Menards Series and NASCAR late model series.

Following the season-ending Xfinity Series race at Homestead-Miami Speedway in 2017, Kennedy stepped out of the cockpit and focused his attention on the day-to-day organization of the sport. And on January 30, 2018, Kennedy was named General Manager, NASCAR Gander Outdoors (Craftsman) Truck Series.

While the Petty family was the first fourth-generation family in American sports history, the France family history still runs deep to this day in NASCAR. At a young age, Kennedy has already had the opportunity to leave a lasting impact on the sport his family has operated for seventy-five years.

He was promoted from that role to Managing Director, Racing Operations and International Development, in 2019. By 2021, he was named Senior Vice President, Strategy & Innovation. In that role, Kennedy was in charge of shaking up the NASCAR Cup Series schedule and innovating it for a modern series—and innovate he did.

On September 14, 2021, NASCAR announced the annual preseason exhibition race traditionally known as The Clash would be held in the Los Angeles Memorial Coliseum on

February 6, 2022. The exhibition event would also serve as the debut of NASCAR's Next Gen car.

For nearly one hundred years, the Los Angeles Memorial Coliseum has been home to some of the greatest sporting events and iconic speeches in American history. It hosted two Olympic Games (1932 and 1984), two NFL Super Bowl games, the 1959 MLB World Series, and has been home to the USC Trojans since 1923. Appearances and speeches at the Coliseum have been made by US Presidents Franklin D. Roosevelt, Dwight Eisenhower, John F. Kennedy, Lyndon Johnson, Richard Nixon, and Ronald Reagan, as well as national and international dignitaries and figures such as Martin Luther King Jr., Pope John Paul II, the Dalai Lama, Nelson Mandela, Billy Graham, and even the Grateful Dead.

Starting in 2022, NASCAR would add its name to that incredible list, but it would take a lot of work to make it happen.

Held at Daytona International Speedway since 1979, Kennedy thought outside the box and took NASCAR to one of the most iconic sports stadiums in the country. In order to hold the event in the home of the USC Trojan football team, NASCAR had to design and construct a racetrack that could sustain a weekend of Cup Series action, put on a good show, and be safe for the drivers behind the wheel and spectators in the stands.

"It's something we frankly have never done before in our sport's history. We had Soldier Field back in 1956 as a statement event, but we've never actually constructed a temporary quarter-mile short track inside a stadium before," said Kennedy.

"I think a lot of internal conversations as you think about logistics, design development of the track, just what that event is truly going to be like," he said. "It's been an opportunity for us to think differently about what that event is going to look like."

In order to plan for the event, Kennedy and his team at NASCAR turned to the sim racing world and iRacing. Working closely with the longtime NASCAR simulation partner, the track digitally built in the Coliseum was modeled after Bowman Gray Stadium, the site of Kennedy's second NASCAR K&N Pro Series East win.

The team then had to go to work building the track in time for the race on February 6, 2022. The USC Trojans played their final home football game on November 27, 2021, and shortly thereafter NASCAR went to work constructing a half-mile asphalt racetrack within the confines of the football stadium.

By the time teams showed up in early February, the facility was completely transformed into a racetrack ready to host NASCAR's biggest stars on one of the biggest stages. Throughout it all, drivers praised the track and NASCAR's efforts to build a quality racing surface in such a short amount of time.

"When I walked in on Thursday through one of those tunnels to take a look at it, it blew me away. Aesthetically this place is beautiful. To know that they did this in thirty days or less, I don't know how they did it," said driver and co-owner Brad Keselowski.

"I think it's awesome; everything about it," said 2021 NASCAR Cup Series champion Kyle Larson.

"Staging outside, driving through the tunnel, getting out there—it all kind of has that grassroots feel to it a little bit. I feel like NASCAR did an amazing job with the racetrack and, so far, I don't really have any comments of concern. I think that's something good because I feel like, a lot of the time, us drivers think we're way smarter than NASCAR and we could have done a better job than them with whatever they were doing. Hats off to NASCAR and what they've done to build a racetrack here."

"Honestly, I think everyone kind of didn't know what to expect when we announced this. When we got here yesterday and were able to walk around and now sitting up here, this is awesome. What a cool venue. A great event, I think," Ryan Blaney said after the weekend's first practice. "The track was really slick getting going this morning, that is for sure. But it got grip by the end of it and everyone learns pretty quickly here. I couldn't tell it was a temporary track. It is asphalt. You can run on it."

Taking place the week before the 2022 NFL Super Bowl—also being held in Los Angeles—and two weeks before the Daytona 500, the Busch Light Clash at the Coliseum had more attention than any other Clash before from both NASCAR and non-endemic media outlets. FOX Sports continuously promoted the event throughout the NFL Playoffs, rivaling that of the Daytona 500.

For the race itself, Kennedy and the team went back to their short-track roots and set the format similar to most NASCAR weekly racing series events.

The 40-car field set qualifying laps and were then divided into heat races. The top-four finishers from each of the four 25-lap Heat Races advanced to the main event. The rest of the field was divided into two 50-lap Last Chance Races, with the top three finishers advancing to the main event. The final spot was reserved for the highest driver in the 2021 point standings that had not already locked into the main event.

In all, twenty-three NASCAR Cup Series cars would take the green flag for the 150-lap main event on February 6, 2022.

The first race ever for the Gen 7 car, there were a host of unknowns throughout the field and NASCAR leadership. While the car had been tested and adjusted for years, it had never been in a true competition setting. Although many teams had concerns about supply chain issues, few available parts, and a lack of cars prepared to start the season, the short-track nature of the Clash meant fenders were likely to be wrinkled and damaged cars headed back to the shop.

Kyle Busch and Tyler Reddick led the 23-car field to the green flag of the 2022 Busch Light Clash at the Coliseum and made history as the first NASCAR race at the iconic facility.

Reddick dominated the first part of the event, which was split into two 75-lap halves with a break in between. However, under caution on Lap 53 of 150, Reddick had issues with the new transaxle and was forced to retire from the race. Chase Briscoe had the same issue on the same lap. Both were examples of the drivers and teams adjusting to the new Next Gen car and its sequential shifter with five gears.

Busch then took command of the race until Joey Logano began to put the pressure on for the top spot. Taking the lead on the final restart of the day, Logano led the final 35 laps to score the victory and write his name in the history books. To top it off, Logano's wife was pregnant with their third child, who was born the following day.

"I can't believe that we're here. LA Coliseum, we got the victory with the old Shell-Pennzoil Mustang," Logano said after climbing from the car. "This is an amazing event. Congratulations, NASCAR. Such a huge step in our industry to be able to do this, put on an amazing race for everybody. I'm out of breath. I'm so excited about this. It was a big win. This is big."

The event marked a six-year ratings high and was seen by 4.28 million viewers on FOX, and the crowd was electric, enthusiastic, and diverse.

"I think the energy behind the crowd here was amazing. I was walking back to the media center right after the event, ton of people smiling, laughing. The energy was really just contagious," said Kennedy. "It was great to see the fan reaction."

On March 10, 2022, Kennedy was named Senior Vice President, Racing Development & Strategy in large part for his role in bringing the Clash to the LA Memorial Coliseum.

"In a short time, Ben Kennedy has proven to be an innovative, thoughtful leader with bold ideas and a tireless work ethic. He returns to his competition roots with this new role, one in which he will no doubt excel," Phelps said in a statement.

The Clash returned to the LA Memorial Coliseum for the 2023 season, with NASCAR once again building a half-mile asphalt racetrack from scratch in nearly no time at all. The promotion was once again second-to-none, and the fans packed the historic facility, which was won by veteran driver Martin Truex Jr.

To promote the second annual event and NASCAR's 75th anniversary season, Richard Petty and up-and-coming African American driver Rajah Caruth were part of a NASCAR-themed float in the 2023 Rose Bowl Parade.

While NASCAR's efforts at the Coliseum have been successful despite the challenges, it has many inside the sport wondering where they can take this concept next. In pulling this off successfully, NASCAR and the entire industry proved it could produce a purpose-built track capable of producing quality racing in a nontraditional facility in a short amount of time.

"As we think about kind of new markets, in particular for some of our international series, I think it opens the door to explore new cities and new markets, emerging markets, that we haven't been to before. Something that we're definitely looking at," said Kennedy. "I think part of the reasoning behind moving to the Coliseum with the multitude of others is really proof of concept: proof of concept for the track, the event, the format, the supporting events, all the things that really go around that weekend."

A radical thought and huge risk. Kennedy never wavered in his determination to think outside the box and take NASCAR places it had never been before. That same way of thinking is how his family built, grew, and sustained NASCAR into the multi-billion-dollar industry it is today. Taking that legacy and facing it head-on, Kennedy seems poised to lead NASCAR into the future much like his great-grandfather and grandfather before him.

NASCAR Takes on the 24 Hours of Le Mans

THE 24 Hours of Le Mans, the iconic sports car race in France, celebrated its 100th year in 2023, and NASCAR has played a small part in the celebration during its 75th year.

A partnership between NASCAR, Chevrolet, Hendrick Motorsports, and Goodyear will see a NASCAR Cup Series style car venture to Europe to compete in a class all its own. The Garage 56 project was announced on March 17, 2022, but the venture was not the first time NASCAR headed overseas to race with the world's elite endurance racers.

In 1976, two NASCAR Cup Series style cars were loaded up and shipped to France to represent NASCAR and America in a newly created Grand International Class. During the United States bicentennial celebration of the signing of the Declaration of Independence, this was motorsports exchange was in part to commemorate the 200th anniversary of France (the country, not the promoter) helping the thirteen American colonies win the Revolutionary War against Great Britain.

Bill France Sr. tapped team owners Junie Dunlavey and Hershel McGriff to prepare the cars and represent the sport overseas. Dunlavey put together a Ford Torino with drivers Dick Brooks, Dick Hutcherson, and Frenchman Marcel Mignot. McGriff prepared a Dodge Charger and drove the car with his son, Doug.

"Before we got there, they'd taken the cars all over France on a flatbed truck," Junie Donlavey said. "People didn't know what to make of them, being so big and painted up and all. They called them 'le grande monster'—or something like that. I never did learn exactly what they were saying."

The cars required some adjustments to conform with the rules of the event as opposed to NASCAR's rules. Windshield wipers, functioning lights, and side mirrors were among the items added to the NASCAR stock cars to race in the 24-hour event.

McGriff's Dodge Charger did not last long in the 24-hour race, retiring after just two laps with engine issues. The Dunlavey team of Brooks, Hutcherson, and Mignot completed 104 laps before exiting the race with a problem with the transmission.

In 2011, Michael Waltrip and Rob Kauffman, then co-owners of Michael Waltrip Racing, competed in the 24 Hours of Le Mans, but did so in a Ferrari 458 GTC, not a NASCAR Cup Series style car. Kauffman returned in 2012, this time with driver Brian Vickers and running a Ferrari 458 Italia GTC.

For the 90th running of the 24 Hours of Le Mans in 2014, Jim France—son of Bill France Sr. and the founder of the Grand Am Series—gave the command to fire engines, but there were no NASCAR-related cars on the iconic grid.

In 2023, that all changes when the Garage 56 Project hits the track in a class of its own. Announced in early 2022, the collaborative project between NASCAR, Chevrolet, Hendrick Motorsports, and Goodyear developed and adapted a NASCAR Cup Series Gen 7-style car.

"From the early days of NASCAR, it was important to my father that we played a visible role in international motorsports, and there is no bigger stage than the 24 Hours of Le Mans," said Jim France, NASCAR Chairman and CEO. "In partnering with

Hendrick Motorsports, Chevrolet, and Goodyear, we have the winningest team, manufacturer, and tire in NASCAR history. We look forward to showcasing the technology in the Next Gen car and putting forward a competitive entry in the historic race."

Ahead of the 2023 NASCAR season, the driver lineup was announced for the historic return to Le Mans. On January 28, NASCAR announced it would entrust its Garage 56 entry to seven-time NASCAR Cup Series champion Jimmie Johnson, 2009 Formula One champion Jenson Button, and 2010 24 Hours of Le Mans winner Mike Rockenfeller. Sports car ace Jordan Taylor was tapped as the team's driving coach and backup driver.

"As a lifelong racing fan, I have always dreamed of racing certain cars, with and against certain drivers and competing in certain events," said Button. "In June, a number of those dreams will come true in one event when I get to bring NASCAR to the world stage alongside my pals Jimmie and 'Rocky' [Rockenfeller] for the 100th anniversary of the most prestigious race in the world. I'm really looking forward to sharing this journey with NASCAR, Hendrick Motorsports, Chevrolet, and Goodyear, and current and future NASCAR fans from around the world."

The Garage 56 car and livery debuted at Daytona International Speedway on February 17, 2023 on the weekend of the Daytona 500, and the drivers held their final test session on April 18, at Sebring International Raceway in Florida. The four drivers ran both the primary and backup cars, worked on adjusting to the pit stops, driver changes, and technical rules that would need to be learned ahead of the June event.

"We've put in many hours, a lot of work, a lot of collaboration and I feel like we're in a really good spot," Johnson said following the final test. "This has been a great test this week, really just working on the small details literally down to things like what color are the buttons on the steering wheel and is the black light inside the car bright enough in the right areas that we can see the dayglow buttons to push at the right time."

While this will mark NASCAR's first time at Le Mans in nearly 50 years, the Garage 56 project is following in the footsteps put in place in 1976 by Bill France Sr. and the group of owners and drivers that first prepared to take NASCAR overseas to compete in one of the world's most prestigious motorsports events.

Works Cited

"2017 Axalta Presents the Pocono 400 Pit Stop Data," https://www.racing
-reference.info/pitstops/2017–14/W/.

ABC Sports broadcast. August 6, 1994.

"A Fresh Start*" Race: Bubba Wallace*, produced by 300 Studios, Boardwalk
Pictures, and NASCAR Productions, Episode 1, February 22, 2022,
Netflix.

A Perfect Storm: 1979 Daytona 500. Directed by Brad Lockhart, NASCAR
Productions, February 13, 2015.

An Interview with: Steve O'Donnell and Ben Kennedy, NASCAR Media
Conference, February 6, 2022.

Assael, Shaun. "A Real Lifesaver," *ESPN The Magazine: Dale Earnhardt 2010
Hall of Fame Collector's Issue,* 2010.

Bailey, Greg. "200 mph . . . Parsons cracks magic mark," *Gadsden Times*. April
30, 1982.

Bell, Mike. "Mrs. Betty Lilly, NASCAR Pioneer," GeorgiaRacingHistory.
com. March 11, 2011. http://georgiaracinghistory.com/2011/03/11/ mrs
-betty-lilly-nascar-pioneer/.

Berggren, Dick. "An interview with Bill France," *Stock Car Racing, Volume
31, Number 3*. Stock Car Racing Publishing, Ltd., New York, New York,
March 1996.

Berggren, Dick. "Jeff Gordon: Winston Cup Superstar?" *Stock Car Racing:
Volume 28, Number 2*. Stock Car Racing Publishing, Ltd. New York, New
York, February 1993.

Bledsoe, Jerry, *The World's Number One, Flat-Our, All-Time Great, Stock Car
Racing Book*. Down Home Press, Asheboro, North Carolina, 1975.

Bourcier, Bones. "Kyle Petty: On the Threshold of Greatness," *Stock Car
Racing Volume 28 Number 8*. Stock Car Racing Publishing, Ltd., New
York, New York, August 1993.

Branham, H.A. and McKim, Buzz. *The NASCAR Vault*. becker&mayer! Books, Bellevue, Washington, 2004.

Branham, H.A. *Bill France Jr.: The Man Who Made NASCAR*. Triumph Books, Chicago, Illinois, 2010.

"Brickyard 400," ABC Sports broadcast. August 6, 1994. http://www.calder park.com.au/history/the-thunderdome.aspx.

Canadianracer.com/nascarincanada.

"Chevy NCS at the L.A. Memorial Coliseum: Kyle Larson and Chase Elliott Press Conference," NASCAR Transcript, February 5, 2022.

"Chip Ganassi sells entire NASCAR team to Trackhouse Racing," *Associated Press*. June 30, 2021.

Crandall, Kelly, "Elliott not holding back on Next Gen safety issues," *Racer*. October 1, 2022. https://racer.com/2022/10/01/elliott-doesnt-hold -back-on-next-gen-safety-issues/.

Crossman, Matt. "An Oral history of the '84 Firecracker 400," *NASCAR Illustrated*. July 2, 2014.

Daniels, J.K. "NASCAR's SuperTrucks: A Monumental Hit," *Stock Car Racing, Volume 31, Number 3*. Stock Car Racing Publishing, Ltd., New York, New York, March 1996.

Darlington Raceway. (2014, August 26). [Press release].

Darlington Raceway. (2015, May 21). [Press release].

Darrell Waltrip media availability, Bristol Motor Speedway, April 17, 2015.

Evernham, Ray. "Winning the Brickyard 400," *Stock Car Racing Volume 29 Number II*. Stock Car Racing Publishing, Ltd., New York, New York, November 1994.

"Fastest NASCAR Cup Series Pit Crews of 2022 Based on Average Four-Tire Stop," December 17, 2022. https://www.nascar.com/gallery/fastest -nascar-cup-series-pit-crews-of-2022-based-on-average-four-tire-stop/.

Fielden, Greg. *Forty Years of Stock Car Racing: The Beginning, 1949–1958*. Galfield Press, Pinehurst, North Carolina, 1988.

Fielden, Greg. *Forty Years Of Stock Car Racing: The Superspeedway Boom, 1959– 1964*. The Galfield Press, Pinehurst, North Carolina, 1988.

Fielden, Greg. *Forty Years of Stock Car Racing: Big Bucks and Boycotts, 1965– 1971*. Galfield Press, Pinehurst, North Carolina, 1988.

Fielden, Greg. *NASCAR: The Complete History*. Publications International, Ltd, Lincolnwood, Illinois, 2009.

"Ford Performance Notes and Quotes – Kevin Harvick and Brad Keselowski," NASCAR Transcript, February 5, 2022.

Ford Racing. (2010, June 28). [Press release].

"Ford Performance Notes and Quotes – Ryan Blaney and Austin Cindric," NASCAR Transcript, February 5, 2022.

"Furniture Row Racing to Cease Operations After 2018 Season." September 4, 2018.

Gelston, Dan. "Dethroned King: Petty hurt as Johnson takes over race team," *Associated Press*. February 18, 2023.

Golenbock, Peter. *American Zoom: Stock Car Racing–from the Dirt Tracks to Daytona*. Macmillan General Reference, New York, New York, 1993.

Golenbock, Peter. *The Last Lap*. MacMillan, New York, New York, 1998.

"Guthrie Finishes in Fifteenth." Statesville Record & Landmark, May 31, 1976, pp. 6B–6B.

Hamlin, Denny, "We're Gonna Get White Hot," *Actions Detrimental with Denny Hamlin*, Dirty Mo Media. April 11, 2023. https://www.dirtymomedia.com/actionsdetrimental/episode/253f324b/were-gonna-get-white-hot.

Harris, Mike. "Petty family goes from glory to tragedy," *Associated Press*. May 12, 2000.

Hembree, Michael. *100 things NASCAR fans should know & do before they die*. Triumph Books LLC. Chicago, Illinois, 2012.

"In His Own Words: Richard Childress, powered by Chevrolet [Part IX]," Richard Childress Racing, September 26, 2019. https://www.rcrracing.com/news-media/videos/richard-childress-dale-earnhardt-kevin-harvick-2001.

Jackson, Mardy. "Crazy Kyle," *Inside NASCAR Magazine, Volume 2, No. 2*. The Quarton Group, Inc., May/June 1998.

Jeff Gordon media availability. August 15, 2014.

Jeff Gordon media availability on February 27, 2015, at Atlanta Motor Speedway.

Kennedy, Ben. Zoom Press Conference Interview, NASCAR media Conference, September 15, 2021.

"L.A. Coliseum TV Interviews with Joey Logano, Kyle Busch, Erik Jones, Austin Dillon – 020622," NASCAR Transcript, February 7, 2022.

Lentati, Sara. "The death that changed NASCAR," *BBC World Service*. April 29, 2015. http://www.bbc.com/news/magazine-32413903

Lewis, Jon, "NASCAR's 'Clash' hits high in L.A. debut, return to FOX," Sports Media Watch, February 9, 2022. https://www.sportsmediawatch .com/2022/02/nascar-clash-ratings-highest-since-2016-fox-los-angeles -coliseum/.

Live Fast Motorsports NASCAR Cup Racing Team. "NASCAR's Newest Cup Series CEO-Jessica McLeod."

Live Fast Motorsports NASCAR Cup Racing Team, 10 Feb. 2022, https://teamlivefast.com/news/f/nascar%E2%80%99s-newest-cup -series-ceo—jessica-mcleod.

Long, Dustin, "Dale Jr. thrilled with the chance to drive Next Gen car at Bowman Gray," NBC Sports, October 26, 2021. https://nascar.nbcsports .com/2021/10/26/dale-jr-tony-stewart-clint-bowyer-thrilled-with -chance-to-drive-next-gen-car-at-bowman-gray/.

McCullough, Bob. *My Greatest Day in NASCAR.* Thomas Dunne Books, St. Martin's Press, New York, New York, 2000.

"Meet Thad," *About Thad Moffitt*, April 2023. https://thadmoffitt.com /about/.

Motorsports on NBC. "Coffee With Kyle: Lesa France Kennedy sheds light on cultural shift in NASCAR," YouTube video, 22:18. July 19, 2021. https: //www.youtube.com/watch?v=h2KpZzbdg-8.

Myers, Bob. "The First World 600," Charlotte Motor Speedway Coca-Cola 600 race program, 2009.

NASCAR. (2005, June 1). [Press release].

NASCAR. (2006, March 6). [Press release].

NASCAR. (2020, June 16). *NASCAR Appoints Brandon Thompson to Vice President, Diversity and Inclusion*, [Press Release].

NASCAR. (2022, November 10). "Best Season Ever? . . . NASCAR Closes Book on Extraordinary 2022," [Press release].

NASCAR. (2022, March 17). "NASCAR, Hendrick Motorsports Announce Pursuit of Garage 56 Entry at Le Mans," [Press release].

NASCAR Hall of Fame induction ceremony. May 23, 2010.

"NASCAR Implements Plan to Welcome Back Guests at Select Tracks," NASCAR press release. June 9, 2020.

NASCAR. Media availability with Austin Dillon, Justin Alexander, Richard Childress. February 18, 2018.

NASCAR. Media availability with Ben Kennedy, Hon. Lori Lightfoot, Steve Phelps, Bubba Wallace, Kara Bachman, Lesa France Kennedy. July 19. 2022.

NASCAR. Media availability with Brian France, Mike Helton, Steve Newmark, Steve O'Donnell, Tommy Baldwin, Brent Dewar, Rob Kauffman, and Richard Petty, February 9, 2016.

NASCAR. Media availability with Chase Elliott. July 4, 2021.

NASCAR. Media availability with Jimmie Johnson and Chad Knaus. March 1, 2015.

NASCAR. Media availability with Steve Phelps, February 12, 2021.

NASCAR. Media availability with Kevin Harvick. May 17, 2020.

NASCAR. Media availability with Steve O'Donnell. May 18, 2020.

NASCAR. Media Availability with Steve O'Donnell and Joie Chitwood III. February 21, 2015.

NASCAR. Media availability with Steve Phelps. February 12, 2021.

NASCAR. Media availability with Steve Phelps. March 17, 2020.

NASCAR. Media availability with Steve Phelps. November 7, 2020.

"NASCAR Names O'Donnell Chief Operation Officer; Kennedy Returns to Competition Team," NASCAR Press Release, March 10, 2022.

"NASCAR President Steve Phelps sees need for Next Gen safety enhancements," Motorsports on NBC, October 9, 2022. https://www.youtube.com/watch?v=17cq6Lklhvc.

NASCAR. Teleconference with Steve O'Donnell and John Bobo. April 30, 2020.

NASCAR NEXTEL Cup Yearbook 2004. UMI Publications, Inc., Charlotte, North Carolina, 2004.

NASCAR NEXTEL Cup Series Yearbook 2006. UMI Publications, Inc., Charlotte, North Carolina, 2006.

NASCAR Winston Cup 1994. UMI Publications, Inc., Charlotte, North Carolina. 1994.

"Next Gen car makes on-track test debut at Richmond," Official NASCAR Release, NASCAR.com. October 9, 2019. https://www.nascar.com/news-media/2019/10/09/nascar-next-gen-car-photos-look-debut-richmond/.

North Wilkesboro Speedway. (2023, September 8). "NASCAR All-Star Race Moves to Historic North Wilkesboro Speedway for 2023," [Press release].

North Wilkesboro Speedway. (2023, March 21). "NASCAR Cup Series Drivers Take on North Wilkesboro in Goodyear Tire Test" [Press release].

O'Malley, J.J. *Daytona 500: 50 Years.* Publications International, Ltd., Lincolnwood, Illinois, 2007.

Orabona, Sandrine, director. The Adventures of Janet Guthrie. YouTube, Fox Sports Films in Association with NASCAR Productions, 2018, https://www.youtube.com/watch?v=MgSBIM6q_qI.

"Our Impact," *Victory Junction.* April 2023. https://victoryjunction.org/our-impact/.

Pearce, Al, "How NASCAR's 'Le Grande Monster' Shook Up the 24 Hours of Le Mans in 1976," *Autoweek.* March 23, 2022. https://www.autoweek.com/racing/nascar/a39517984/nascar-le-grande-monster-shook-up-24-hours-le-mans-1976/

Phelps, Steve, "State of the Sport address with Steve Phelps & Steve O'Donnell," NASCAR Media Conference, November 4, 2022.

Pierce, Daniel S. *Real NASCAR: White Lightning, Red Clay, and Big Bill France.* The University of North Carolina Press, 2010.

Pockrass, Bob [@bobpockrass], "Richard Childress gives a status update on how many Next Gen cars the team has and whether he is nervous about having enough cars. He added later that he hoped to have 10 total cars (five for each of his teams) before the West Coast swing in late February." *Twitter,* January 11, 2022. https://twitter.com/bobpockrass/status/1481006868944371723.

Ricky Stenhouse, Jr. media availability. February 19, 2023.

"SHR Post-Race Recap: Darlington," Stewart-Haas Racing. September 4, 2022. https://www.stewarthaasracing.com/shr-post-race-recap-darlington2-2022/.

Smothers, Jimmy. "No one surprised by Parsons' effort," *Gadsden Times.* April 30, 1982.

Spencer, Lee, "Denny Hamlin calls for a complete redesign of the Next Gen car," RacinBoys Broadcasting Network, October 1, 2022. https://racinboysscoop.com/news/news.asp?NewsID=89700.

Sports Illustrated Presents: Winston Cup 2001. Time Inc., 2002.

Stern, Adam, "NASCAR charters doubling in value since last year," *Sports Business Journal,* Street & Smith's. November 2, 2021.

Stern, Adam, "Sources: NASCAR charters going for $20M despite industry skepticism," *Sports Business Journal,* Street & Smith's. October 10, 2022.

Steve Phelps media availability. June 25, 2020.

Sturniolo, Zach, "NASCAR enhances safety of Next Gen car, mandates data collection for 2023," NASCAR.com. February 1, 2023. https://www .nascar.com/news-media/2023/02/01/nascar-enhances-safety-of-next -gen-car-mandates-data-collection-for-2023/.

Team Chevrolet. (2015, March 1). *NASCAR Sprint Cup Series Folds of Honor QuikTrip 500 Atlanta Motor Speedway Team Chevy Race Notes and Quotes* [Press Release].

"The Day: 1992 Hooters 500," NASCAR Media Group. 2011.

"The Day: Richard Petty's 200th Win," NASCAR Media Group. June 30, 2011.

Thompson, Neal. *Driving With The Devil*. Crown Publishers, New York, 2006.

Waid, Steve. "At Last: Earnhardt finally bags the big one," *NASCAR Winston Cup Scene Volume XXI No. 40*. Charlotte, North Carolina, February 19, 1998.

Waid, Steve. "Behind Closed Doors: How NASCAR Came To Indianapolis," July 18, 2014. http://www.popularspeed.com/behind-closed-doors-how -nascar-came-to-indianapolis/.

Wallace, Bubba, "Bubba Wallace responds after FBI hate crime investigation," *Don Lemon Tonight*, hosted by Don Lemon, CNN, June 24, 2020.

Wallace Jr., Darrell [@BubbaWallace]. "Shits getting old.. hell its been old. Wtf is gonna change?! #GeorgeFloyd." *Twitter*, May 27, 2020, https: //twitter.com/BubbaWallace/status/1265649929701244928.

Winston Cup Illustrated: 1992 Season Review. Griggs Publishing Company, Inc., Concord, North Carolina, January 1993.

"Wood Brothers Appearance with 1965 Indy-Winning Lotus at Goodwood Festival of Speed," June 29, 2010, http://woodbrothersracing.com /classic-memories/wood-brothers-appearance-with-1965-indy-winning -lotus-at-goodwood-festival-of-speed/.

"Yarborough recalls historic first night race win at BMS," June 6, 2011. http://www.bristolmotorspeedway.com/news_media/news_releases /yarborough-recalls-historic-first-night-race-win-bms-590764.html.

Zeller, Bob. "Between a Cathouse and a Dog Track." *Car and Driver*, March 2008.

Interviews with the author include: Ray Evernham, Ned Jarrett, Buz McKim, Robin Pemberton, Jack Roush, Ken Squier, Russ Truelove, H. A. "Humpy" Wheeler, Frank Scott, Derrell Edwards, and Leonard Wood.

Index

Author Bio

Jay W. Pennell has been a part of the NASCAR industry in multiple facets for over fifteen years. His career path has taken him to many parts of the sport; from independent journalist, to social and digital marketing, to team public relations, to NASCAR gaming.

Pennell has written for outlets such as FOX Sports, NASCAR.com, *Playboy*, and *Athlon Sports*, and was one of NASCAR's first citizen journalists. He also worked with Richard Childress Racing, handling public relations, content creation and partnership management for four seasons.

While with RCR, he celebrated many wins, including the 2018 Daytona 500 and the 2019 NASCAR Xfinity Series championship, helped curate the organization's 50th anniversary content strategy, and worked with drivers such as Austin and Ty Dillon, Brendan Gaughan, Daniel Hemric, Ben Kennedy, Brandon Jones, Paul Menard, Ryan Newman, and Tyler Reddick.

He also served as the General Manager of the eNASCAR Heat Pro League, later moving to Brand Manager, NASCAR and INDYCAR with Motorsport Games. Pennell currently contributes to outlets such as *Racing America* and helps young up-and-coming drivers with marketing and partnership guidance.